The Wireless Internet Explained

Related Titles from Digital Press

RELATED TITLES

Tony Redmond, *Microsoft® Exchange Server for Windows 2000: Planning, Design, and Implementation,* ISBN 1-55558-224-9, 1072pp

Kieran McCorry, *Connecting Microsoft® Exchange Server,* ISBN 1-55558-204-4, 408pp

Sue Mosher, *The Microsoft® Outlook® 2000 E-mail and Fax Guide,* ISBN 1-55558-235-4, 512pp

Jerry Cochran, *Mission-Critical Microsoft® Exchange 2000: Building Highly Available Messaging and Knowledge Management Systems,* ISBN 1-55558-233-8, 320pp

Micky Balladelli and Jan De Clercq, *Mission-Critical Active Directory™: Architecting a Secure and Scalable Infrastructure,* ISBN 1-55558-240-0, 512pp

Mike Daugherty, *Monitoring and Managing Microsoft® Exchange 2000 Server,* ISBN 1-55558-232-X, 432pp

For more information on or to order these and other Digital Press titles please visit our website at www.bh.com/digitalpress!

At www.bh.com/digitalpress you can:

- Join the Digital Press Email Service and have news about our books delivered right to your desktop
- Read the latest news on titles
- Sample chapters on featured titles for free
- Question our expert authors and editors
- Download free software to accompany select texts

The Wireless Internet Explained

John Rhoton

Digital Press
An imprint of Butterworth-Heinemann

Boston • Oxford • Auckland • Johannesburg • Melbourne • New Delhi

Copyright © 2002 Butterworth–Heinemann

 A member of the Reed Elsevier group

All rights reserved.

Digital Press™ is an imprint of Butterworth–Heinemann.

All trademarks found herein are property of their respective owners.

No part of this publication may be reproduced, stored in a retrieval system, or transmitted in any form or by any means, electronic, mechanical, photocopying, recording, or otherwise, without the prior written permission of the publisher.

∞ Recognizing the importance of preserving what has been written, Butterworth–Heinemann prints its books on acid-free paper whenever possible.

Library of Congress Cataloging-in-Publication Data

Rhoton, John.
   The wireless Internet explained / John Rhoton.
     p. cm.
   ISBN 1-55558-257-5 (pbk. : alk. paper)
    1. Personal communication service systems. 2. Internet. 3. Computer networks.
I. Title.

TK5103.485.R52 2001
004.67'8--dc21
                                            2001047692

British Library Cataloging-in-Publication Data

A catalogue record for this book is available from the British Library.

The publisher offers special discounts on bulk orders of this book.
For information, please contact:

Manager of Special Sales
Butterworth–Heinemann
225 Wildwood Avenue
Woburn, MA 01801-2041
Tel: 781-904-2500
Fax: 781-904-2620

For information on all Butterworth–Heinemann publications available, contact our World Wide Web home page at: http://www.bh.com.

10 9 8 7 6 5 4 3 2 1

Printed in the United States of America

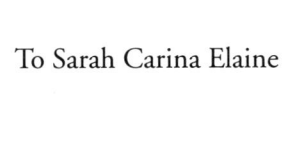

To Sarah Carina Elaine

Contents

Preface — xi

Acknowledgments — xv

1 Wireless Technology — 1

    1.1 Introduction — 1
    1.2 Mobile and wireless — 2
    1.3 The world before wires — 3
    1.4 Wireline communication — 4
    1.5 Emergence of wireless — 6
    1.6 Why wireless? — 8
    1.7 Wireless physics — 10
    1.8 Radio transmission — 17
    1.9 Wireless network architecture — 20
    1.10 What are the implications of wireless? — 22
    1.11 Summary — 23

2 Air Interfaces — 25

    2.1 Air ownership — 25
    2.2 Licensing airwaves — 26
    2.3 Ownership and billing — 28
    2.4 Interface characteristics — 28
    2.5 Current interfaces — 38
    2.6 Wireless PANs — 39
    2.7 Wireless LANs — 41
    2.8 Wireless MANs — 45
    2.9 Wireless WANs — 46
    2.10 Satellite communications — 53
    2.11 Summary — 55

3 Mobile Devices — 59

- 3.1 Device types — 60
- 3.2 Platforms — 63
- 3.3 Applications — 66
- 3.4 Synchronization — 70
- 3.5 Extensibility — 73
- 3.6 Smart cards — 74
- 3.7 Voice processing — 75
- 3.8 Management — 78
- 3.9 Trends in mobile devices — 81
- 3.10 Summary — 82

4 Infrastructure — 85

- 4.1 Networks — 87
- 4.2 Frameworks — 91
- 4.3 SMS—Short Message Service — 92
- 4.4 WAP — 96
- 4.5 i-mode — 104
- 4.6 IrDA — 107
- 4.7 Bluetooth — 108
- 4.8 Pure IP networks — 114
- 4.9 Summary — 115

5 Applications — 117

- 5.1 Why are mobile applications different from traditional applications? — 117
- 5.2 Evolution of applications — 118
- 5.3 Enabling applications for wireless access — 121
- 5.4 Consumer applications — 130
- 5.5 Enterprise applications — 132
- 5.6 Vertical applications — 144
- 5.7 Optimized wireless applications — 145
- 5.8 Increasing the value — 147
- 5.9 Summary — 149

6 Security — 151

6.1	Device security	151
6.2	Air security	156
6.3	Supplementary security	166
6.4	Enterprise requirements	171
6.5	Secure transactions	172
6.6	Summary	174

7 Implementation — 177

7.1	Architecture/design	177
7.2	Development	198
7.3	Management	203
7.4	Summary	204

8 Future Directions — 207

8.1	Mobile architectures	207
8.2	Air interfaces	210
8.3	Networks	217
8.4	Presentation	224
8.5	Development tools	225
8.6	Applications	226
8.7	Pervasive use	229
8.8	Summary	230

List of Acronyms — 233

Index — 237

Preface

Until about two years ago my technical focus was on building messaging and networking configurations for Compaq's international enterprise customers. One of the biggest trends that became apparent to me was the increasing mobility of the workforce. Not only were our users travelling between corporate offices and customer sites, but many had the need to move freely within the buildings too.

At that point I decided that the area was compelling enough to deserve much more focused attention and made an effort to investigate the technology and all the areas that it touches on. This objective was much more ambitious than I had anticipated as I found that almost all applications and infrastructure in the IT business can glean some benefit from mobility.

I don't think anyone really likes wires. They tangle, restrict movement and provide a unique opportunity to trip if you don't keep your eyes glued to the floor. Until recently they have been tolerated out of necessity. But with improvements in battery life and radio frequency techniques many of these wires are no longer needed.

Wireless technology is a fascinating area to watch. It is developing at a breathtaking pace with new standards being proposed and implemented regularly. Just keeping track of the latest status is a great challenge, but a good understanding of the field also means delving into the applicability and interrelationships of the protocols.

My objective in writing this book is to help clarify how some of these standards fit together to produce a productive and efficient solution that adds value to the existing electronic infrastructure. I hope that as you read, you too, will see how exciting and useful some of these new technologies can be.

Audience

The primary audience for this book is readers who want or need to implement wireless technology in their organizations. This will include network administrators who need to extend the existing configurations based on copper wires and optical fibers to a wireless environment. It will also include developers and implementers of applications that need to be enhanced for user mobility.

But beyond these narrow segments I intend the material to be relevant to anyone in the IT industry with an interest in emerging fields. With the wide reach and applicability of wireless and mobility, most professionals will be affected to some degree and need to be equipped with the knowledge to leverage the newest developments.

Structure

I have attempted to take a practical look at wireless technology, without all the formulas only radio technicians can understand. Instead I want to explain the concepts behind the physics and try to provide an overview that clarifies the convoluted set of standards heaped together under the umbrella of wireless.

The book expands on these technical foundations to give a panorama of the crowded landscape of wireless product offerings. It then describes the actual implementation with topics ranging from the selection and deployment of mobile devices to the extremely sensitive subject of security.

While the chapters of the book are largely independent of each other, they will be most readable if approached sequentially from beginning to end.

Chapter 1 gives an overview of how wireless transmissions take place and what the implications of wireless technology are on providing end-to-end connectivity.

Chapter 2 describes some of the many wireless interfaces that are available and explains the trade-offs between them.

Chapter 3 looks at the spectrum of mobile devices and the applications that are currently available on the most common platforms.

Chapter 4 analyzes the existing wireless infrastructure, both in terms of the network topologies and the supporting frameworks including i-mode, WAP and Bluetooth.

Chapter 5 introduces an array of options for enhancing applications to include mobile support and indicates the pros and cons of each.

Chapter 6 broaches the sensitive, but critically important, subject of security and the many areas of wireless technology that must be secured.

Chapter 7 pulls all the technological considerations together and provides guidance on building the complete end-to-end solution.

Chapter 8 closes off the book with an overview of some of the emerging trends in wireless technology and an assessment of their potential impact on future mobile applications.

Feedback

If you have any questions about *The Wireless Internet Explained* or would like to let me know what you thought of it, please don't hesitate to send me mail. At the time of writing my ever-changing e-mail address was John.Rhoton@compaq.com. I would love to hear from you.

Acknowledgments

There is an abundance of information on wireless transmissions available on the Internet which I have drawn on frequently in developing ideas and content. Where possible I have noted the URLs at the end of each of the chapters and I would encourage you to look them up if you would like more indepth information than I have provided. Clearly, these resources are beyond my control so there is some chance that they will not be available when you access them. Nonetheless I hope that most of them will still be active or will at least be easily locatable through a search engine.

I would like to thank Bob Fleischer and Jeff Yaplee for their help in reviewing this book and offering valuable suggestions for improvement. Many of my colleagues in the Emerging Technologies group, the Wireless and Networking Practice and the Wireless Roadmap Team helped me indirectly by providing insightful and creative perspectives on the technologies involved. My management, in particular Mike Travis and Tony Redmond, has been very active in encouraging technical writing which has helped motivate me to keep working.

I would also like to recognize Pam Chester and Theron Shreve at Digital Press, who worked diligently with me through the entire development process. Alan Rose and Lauralee Reinke did an excellent job of ensuring the book was meticulously copyedited and produced to professional standards.

Claudia, my wife, and Sarah, my daughter, did not directly contribute to the content but I am grateful to them for accepting the fact that I was often distracted and devoted less time and attention to them over the past months. They are both wonderful and I am lucky to be part of their family.

Wireless Technology

1.1 Introduction

It is hard to escape the current excitement regarding wireless technology. From the executive who reads his or her e-mail in the airport lounge, to the teenager playing with a new mobile phone, you can see it everywhere you look.

Companies are still restructuring their business processes in order to enter the Internet era. But they realize that advances in technology do not come to an end with the Internet. Already many of them are ensuring that their Web content is accessible to the various mobile devices appearing on the market.

And yet, even though wireless communication has received renewed attention recently, it is nothing new. Smoke signals, human speech, and visual gestures are merely some of the many forms of communication that people have used since the beginning of recorded history. And, of course, all of these were very effective without using any wires whatsoever.

Paradoxically, it was the introduction of wires and their use as a medium for communication that enabled the informational infrastructure of the twentieth century. It was this pivotal advance in technology that provided a framework for the new economy to flourish.

Yet when we speak about wireless technology today, we don't usually mean all technology that does not use wires. We are typically referring to a very specific technique of transmitting electromagnetic energy through the atmosphere.

This type of wireless transmission has been in use most of the century. However, its application has been limited to a small subset of all communication, such as radio and television. Today's increased demand for mobility, along with technological advances in wireless capability, have prompted a crucial change in its use and dramatically expanded the set of potential applications.

This chapter begins with the history of communication techniques and the technologies supporting them. It then gives a high-level perspective on the physical processes involved in wireless transmissions. It concludes by describing how these are integrated into a network to provide complete end-to-end connectivity.

1.2 Mobile and wireless

While mobile access to the Internet is very often associated with wireless technology, it is important to note that the two concepts are actually distinct. There are a number of ways you can be mobile while still using a fixed line, as follows:

- A mobile device (e.g., laptop) can be connected to multiple fixed access points. This would be the case with a traveling salesperson who calls up e-mail from the hotel every evening.

- A person may be mobile (roaming user) and log into multiple fixed devices (PCs) connected with fixed lines. Whether they are stations in an Internet café or guest terminals in a company's reception area, there are numerous opportunities to access the Internet.

It is also possible to have a wireless connection that does not support mobility. The typical reason for this would be if the devices using the connection are too large or heavy (desktop PCs or servers) to be easily transported. It might seem counter intuitive to configure these machines with wireless interfaces. However, the advantage of wireless technology in these scenarios derives from economical considerations such as deployment costs rather than improved accessibility.

Some fixed wireless connections might include the following:

- Wireless Metropolitan Area Networks such as LMDS or MMDS provide last-mile coverage to remote areas.

- Wireless Local Area Networks are being implemented in many corporations as an alternative to laying cable. While many of the stations may be mobile, it is also quite possible to equip stationary desktops with wireless interfaces rather than running the wire to the computer.

One way to think of the distinction between mobile and wireless is that increased mobility is the main incentive for wireless technology. It is the objective. Wireless technology is not an end in and of itself. It is employed as a means either to improve mobility and productivity or to reduce deployment costs.

1.3 The world before wires

The objective of this book is not to provide a comprehensive theory of communications or even to approach an exhaustive history of it. These are interesting, but only peripheral, topics to wireless transmissions.

Nonetheless, there is value in taking a step back and looking at some of the areas related to the technology. In addition to the analogies that we can draw between different techniques of communication, the bordering domains also provide a context for our discussion.

1.3.1 The distance barrier

The biggest barrier to interpersonal communications has always been distance. As long as mankind has existed, people have been able to interact with each other when they were together. At birth we begin to acquire the customs and languages of those around us so that, by talking and pointing, we can express ourselves to anyone in our immediate vicinity.

However, once we leave visual and auditory proximity, our innate capabilities are insufficient. Unless someone can see or hear us, no one will know what we need or want. Clearly this was a problem even in prehistoric times. Some agricultural tribes might have been stationary and able to remain close together. But nomadic peoples or those who lived from hunting and gathering were much more likely to be dispersed and needed to be able to regroup when necessary.

Where there is a need, someone will usually find a solution. And this is no different. Some of the earliest forms of short to medium range communications included smoke signals and drum beats. These were relatively easy to produce; however, even they required a code, or protocol, which had to be invented without any precedent. Once simple codes became prevalent it was no great feat to extend the complexity. But there was a small limit on the amount of information that could be transmitted in any reasonable amount of time. We would probably say today that the bandwidth was very small.

One of the biggest drivers of early over-the-air, long-distance communications was navigation. As trade took to the waters, the maritime industry flourished. Ships ventured greater distances, and it was not feasible for a captain to know the shape of every port or estuary. However, it was also not possible for land-based assistance to give directions until the captain had arrived. Sophisticated systems of torches and lanterns by night, and flags by day, helped to bridge this gap and enabled port authorities to safely guide foreign boats even during poor weather conditions.

1.3.2 Recording information

The ability to transfer information across a spatial expanse is a very important goal of communication. It may sound unnecessarily metaphysical to express it this way, but one important side effect is also the transfer of information across time. In order to do this effectively the data must be recorded, or transferred, using a permanent, rather than transient medium.

Some of the earliest recording systems are cave drawings, inscribed many thousands of years ago. These pictorial systems were not intended for geographical distribution. It is, after all, very difficult to send a cave to the neighboring village. Their purpose was either merely a form of expression or an attempt to preserve ideas for posterity.

Later more mobile media, such as clay tablets, and then papyrus, came onto the scene. With these the first forms of true long-distance communication became possible. By writing messages on parchment it was possible to transport them anywhere in the world.

The security and reliability of this transport depended on the messenger. If the contents were of a sensitive nature, it was advisable to use a personal trusted messenger. However, this was necessarily an efficient approach for the prolific writers who had multiple messages going in different directions at the same time.

Carrier pigeons offered an attractive alternative in terms of speed and reliability. However, they could only service a small set of destinations and were not suitable for ad hoc messages.

In the middle ages messenger services became popular. Their reputation and area of geographical coverage grew until local governments eventually took them up. The result evolved into what is now a global postal service.

1.4 Wireline communication

While the postal service was able to cover virtually any distance, it did have one major drawback. It was slow. The snail-mail we know now operates at lightning speed compared to what our great-grandparents endured in the middle of the nineteenth century. Letters could take months to arrive by train and ship to remote and isolated destinations. Needless to say, the concept of urgency was not yet quite as pronounced as it is today.

1.4.1 Telegraph

The general population, of course, did not appreciate the delays. But they simply had no choice and could hardly imagine a faster medium. That was the situation in 1837 when Samuel Morse demonstrated a telegraph system that had the capability to send and receive electrical signals over copper wires.

The idea caught on quickly. Operators at both ends were able to code messages (using Morse code) into a sequence of on and off signals that could be transmitted by tapping on an electrical switch. At the other end the clicks could be heard and transcribed into writing. Within moments it was possible to send messages over large geographical distances.

The impact of the technology was revolutionary. It only took a few decades for large telegraphic networks to be established both in North America and Europe with regular transatlantic telegraph service.

The most obvious disadvantage of the telegraph was the need to manually code and decode each message that was transmitted. While the speed of transmission was incredible, the speed of entry and decoding was limited by the capabilities of the operators, who had to be particularly skilled in coding and decoding Morse code.

The process was made a little more user-friendly when telex systems automated the coding process. Rather than an electrical on-off switch the operator could use the familiar typewriter keyboard to enter the letters. And at the other end the text was printed and could be given directly to the recipient without manually decoding it.

There were other improvements in telegraphy too. A service called Wirephoto was able to transmit pictures over telegraph lines. This was particularly useful for newspapers, which needed to transfer live photographs of current events to be included in the daily issue of the paper.

But clearly, the most fundamental advance was the invention of the telephone. With it, most other telegraphy systems were reduced to very specific needs.

1.4.2 Telephone

In 1876 Alexander Graham Bell patented an improvement in telegraphy that made it possible to transmit the human voice (rather than only electrical clicks) over copper wires. Even though many early proponents thought the telephone might find its best use as a broadcast medium, it initially worked in pairs as point-to-point connections. Very soon switchboards were established that enabled communication between any two subscribers.

The telephone system has clearly been at the heart of modern telecommunications. Its ubiquitous reach made it the ideal platform for data communications and consequently it is the most common "last mile" of the pervasive Internet.

In over 100 years it has undergone many transformations, including improved switching systems, digitization, and integrated voice and data services (ISDN). While it began with only simple copper wires, many of its trunks now use high-speed optical fibers.

In most countries the telephone networks were initially monopolies, often run by the state. Deregulation has produced an increasing number of competitive telephone offerings. But the Public Switched Telephone Network (PSTN) is facing competition from other sources.

The Internet has made it possible to reach a growing number of people without using the telephone system. Users of cable television links and satellite dishes are able to connect to the Internet through alternate means. As telephone-equivalent functionality (such as voice) becomes increasingly available over Internet Protocol (IP), the telephone network faces new challenges. It is unlikely that it will fade into oblivion but it must change to adapt to the increasingly diverse needs of its users.

1.5 Emergence of wireless

The beginning of wireless transmission was not much later than the beginning of wireline communications. In 1895 Guglielmo Marconi invented the wireless radio, a device that could send and receive a signal at a distance of almost three kilometers by using simple on-off signals similar to the telegraph.

As with the evolution of telegraphy and telephony, there was a need for analog, rather than digital, transmission through the air. Six years later, in 1901, Reginald Fessenden patented an alternator that used continuous waves rather than spark-gap signals and could therefore carry the human voice. Later improvements in the technique used crystals, which received clearer signals, and tubes, which amplified the received signals.

The technology never stopped advancing. Not only did the tuners become more sensitive, but new techniques helped to make major improvements. Early radio transmissions used only amplitude modulation to superimpose audio signals onto radio waves. The switch to frequency modulation reduced the distortion and was able to improve the clarity of reception.

In the 1920s the first attempts to transmit visual images wirelessly were made. The task was more complex than voice transmission, since it involved passing more than one data point at the same time. A sound transmission only tracks one variable (the acoustic amplitude), which can easily be sent across one carrier.

However, pictures consist of shapes that vary in two dimensions as well as time. It is very nearly impossible to impose an analog visual image onto a waveform. Fortunately our visual systems do not require a completely accurate rendition of a visual image in order to interpret it; a fine-grained approximation is sufficient.

The visual transmission, therefore, involves breaking the pictures down into a set of dots and transmitting a sequence of samples. The dimensions of the picture and the transmission scheme are arbitrary, which is the reason that there are several television standards (NTSC, PAL, and SECAM, as well as their variants) in operation, while radios interoperate very well throughout the world.

Nonetheless, the demand for picture transmission far exceeded any of the challenges it faced. The first sets were offered for sale within a decade of their invention. And today they represent one of the most common household appliances in almost every country.

1.5.1 Domain of wireless

Throughout most of the twentieth century there was a clear distinction between the domains of wireless and wired communications. They each reflected the physical nature of the medium, as follows:

- Wire transmissions were generally point to point between two individuals. This mirrors the essence of a copper cable, which has two end points and transmits information between them.

- Wireless communication was primarily limited to broadcast (i.e., radio and television). In this case the medium is air, which is shared by the entire population. Everyone who is in the general area can receive any transmission.

1.5.2 Point-to-point wireless connectivity

Some of the early champions of point-to-point wireless connectivity were police departments, which needed a communications mechanism in order to keep in touch with mobile officers. They began to pilot the technology in the 1920s and were using it extensively by the 1940s.

The barrier to more widespread adoption of wireless communications was the limitation of radio frequencies, which would quickly fill up in populated areas. In 1947 the United States Federal Communications Commission (FCC) began to grant some spectrum to the mobile phones.

But it wasn't until the 1970s that service areas were divided into geographical cells, permitting significant populations to simultaneously transmit over the same frequencies, using the now outdated, but still used, analog system called Advanced Mobile Phone Service (AMPS).

1.6 Why wireless?

During the last few years wireless has emerged from being only a broadcast technology. Instead its use has expanded into the world of personal communications. The first devices to break this barrier on a large scale were pagers.

Pagers didn't typically communicate much information, at least in the beginning. But they had a very important advantage over any fixed device. They were always accessible. In fact, that was really the only functionality they originally provided. It became possible to reach people wherever they were. And this in turn removed the constraint for a person, who needed to be continuously reachable, to remain in one place. Wireless technology opened the door to mobility.

Independent of the evolution of wireless, fixed-line communications have also been advancing by leaps and bounds over the past decades. The culmination of these developments over the past decade has been the Internet. It has ceased to be an esoteric tool used only by a subset of technologists and now has become a household commodity enjoyed by the masses. However, while this evolution is not yet complete, there are limits to how much further it can continue in its present form.

Two notable constraints include penetration and accessibility.

1.6.1 Penetration

This term is used to denote the percentage of the population who own and use a particular technology. On one hand it is theoretically possible for the penetration to exceed one hundred percent, since some people may operate more than one system. However, this is balanced by the fact that for various reasons no technology is adopted by the whole population.

The penetration of the Internet on desktop devices is currently limited due to the following:

- The cost of a desktop is not affordable for large segments of the world's population. Even those who can afford it may not feel the value of the Internet is high enough to justify the investment.
- Many do not live or work in an environment where a computer would be useful. Those who live in very small apartments may prefer to use the space for other purposes. And manual laborers will not benefit greatly from a computer at work.

Wireless devices have an advantage in that they are typically smaller and cheaper than desktop systems. They are, therefore, both more affordable and more easily deployed in constrained environments.

1.6.2 Accessibility

Those who do have a computer are not always near it or using it. Portable computers may be used remotely. But unless they use a wireless connection, they will be very limited in what they can do with regard to the Internet.

Utilization of portable wireless devices can be much higher than for fixed terminals. A miniature appliance that is carried throughout the day and has constant connectivity to the Internet has the potential for continuous use.

Beyond lifting these constraints wireless devices have additional advantages, as follows:

- Comfort—The fact that wireless terminals do not require any connections makes the configuration easier (users do not want to plug and play—they just want to play), particularly when the device is portable. Also, most users perceive the absence of cabling as an advantage in use (the user can move about the room or office while using the device).

- Security—Security is a complex topic, which will be covered in more depth later in this book. Mobile devices are exposed to risks different from stationary systems. While some of these risks represent threats if not properly dealt with, a completely secured wireless solution offers many advantages over its wired counterparts.

In contrast to fixed terminals, which are often shared by multiple users and are exposed when the user is absent, mobile devices are typically of a very personal nature. They are kept near the user at all times and are rarely shared. This, in and of itself, makes them more secure.

1.7 Wireless physics

Radiation has an ominous sound to it. We often associate the word with some of its more destructive forms, such as nuclear blasts and accidents in atomic power plants. However, it is not altogether an evil force. We would never be able to survive, for example, without the sun's heat and light, which are also transferred as radiation.

1.7.1 Are radio frequencies a health risk?

With any new technology, particularly one that uses a propagation technique that can penetrate our bodies, it is legitimate to ask what the potential damage can be to our health. There have been many claims that mobile phones may have a carcinogenic effect, specifically on the brain. While these claims are heavily contested and have been refuted by counterclaims, the unfortunate conclusion one must draw from looking at the debate is that it is not possible to make an irrefutable case: neither that it is harmless nor that it is dangerous.

After several decades of operation, we can conclude that radio and television broadcasting, using similar frequencies, is not a significant risk, but it also operates at greater distances from the receiving users. In time, we will certainly find out whether mobile phones do cause any damage after years of prolonged use. But in the interim we can only optimistically assume that the reward is greater than the risk.

1.7.2 Radiation

Radiation is merely the process of transmitting energy through space. It can be in the form of waves or particles. While mechanical radiation is transmitted as waves only through matter, electromagnetic radiation, used by wireless technologies, is independent of matter (although its speed and direction are affected by matter in its path).

Electromagnetic radiation is produced by the oscillation or acceleration of an electric charge. As its name implies, the waves have both electric and magnetic components. Each oscillation involves a displacement of both electric and magnetic fields of force in space.

There is an inverse relationship between the frequency and the wavelength. This is very logical once you think about it. Electromagnetic waves travel at the speed of light (3×10^8 m/sec—in a vacuum). This means that the theoretical distance covered by radiation in one second is 3×10^8 m. If this distance is composed of n waveforms, then the length of each waveform must be 3×10^8 m divided by n.

Another way to express the relationship is with the following formula:

Frequency × Wavelength = Speed of light

So, for example, a wavelength of 30 m would correspond to 10 MHz (10 million cycles per second). Other values are given in the following chart.

1 mm	300 GHz	1 m	300 MHz	1 km	300 KHz
3 mm	100 GHz	3 m	100 MHz	3 km	100 KHz
10 mm	30 GHz	10 m	30 MHz	10 km	30 KHz
30 mm	10 GHz	30 m	10 MHz	30 km	10 KHz
100 mm	3 GHz	100 m	3 MHz	100 km	3 KHz
300 mm	1 GHz	300 m	1 MHz	300 km	1 KHz

There are fundamentally two different directions in which wave oscillations can occur. Longitudinal waves, such as sound waves, displace matter in the direction of the wave itself. Each wave means that the air is successively compressed and then released, similar to a push and pull movement. The wave is propagated outward from the source and the wave motion (the compression and rarefaction) is active in the same direction.

Electromagnetic waves have a different type of wave motion. They are called transverse, which means that the vibrations are orthogonal to the direction of motion. This would be similar to the motion of a guitar string. When it is released at one end, the wave propagates along the string, but the wave motion is perpendicular to the string.

Figure 1.1
Reflection.

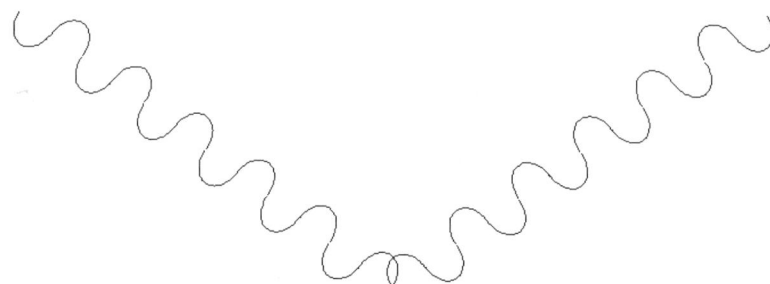

In a vacuum the wave motion would continue infinitely in the same direction without changing characteristics. However, our physical universe is not a vacuum, and any energy will eventually encounter matter or other forms of energy.

When radiation strikes a body of matter, there are several different processes that take place.

Reflection

A waveform is reflected when its path changes direction as it collides with matter. Rather than entering the body of matter, it bounces off the surface and continues on, whereby the angle of reflection is identical (but opposite) to the angle of incidence. (See Figure 1.1.)

Refraction

A refracted waveform does enter the body of matter. Depending on the density of the body being entered and the body from which it arrives, the angle will change, causing a slight bend in the path of the wave. (See Figure 1.2.)

Figure 1.2
Refraction.

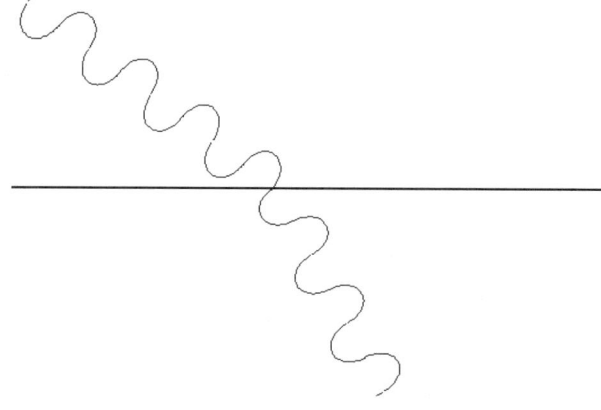

Figure 1.3
Attenuation.

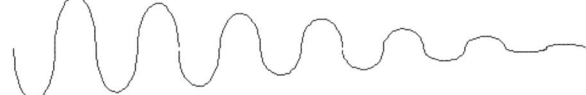

Attenuation

Attenuation is the reduction of signal strength over distance due to conversion of energy to other forms, such as heat. (See Figure 1.3.)

Absorption

Some bodies absorb a part of the wave or even the entire wave. Consequently, less of the wave is able to be reflected or is passed through refractively. (See Figure 1.4.)

Interference

Interference is the term used when a waveform collides with another source of energy, either at the same frequency or in a neighboring band. It is even possible for a waveform to interfere with itself when it reflects off different objects. This is called multi-path interference. (See Figure 1.5.)

Figure 1.4
Absorption.

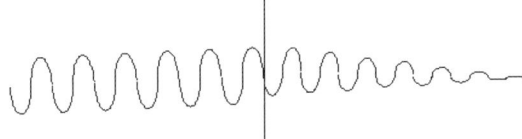

Figure 1.5
Interference.

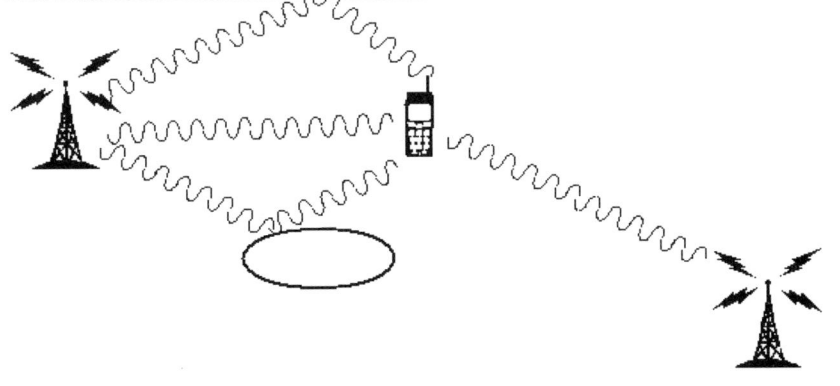

1.7.3 Sound

Sound is also a wireless medium of communication in the most literal sense. Similar to electromagnetic radiation, it also propagates using wave motion. However, these waves are not electric or magnetic, but mechanical. The vibrations are considered sound when they occur in the range of 15 Hz to 20 KHz.

The term *sound* is often also used to include liquid and solid vibrations, but most typically it refers to successive compressions and rarefactions of air. It travels at a speed of 332 m/sec through dry air at 0° Celsius. The movement is generally in a straight line, although it is subject to both reflection and refraction. As it travels, it loses its intensity at a rate proportional to the square of the distance it has traveled.

While our mind is not explicitly aware of the actual waves that compose sound, we are able to discern three components: Pitch (frequency), intensity (amplitude), and timbre (harmonic constitution) of the waveforms.

Sound is the medium used by the most common form of interpersonal communication: Speech. When speaking, the vocal chords vibrate, generating sound waves, which compress the air in cycles. These vibrations occur in the direction of the wave motion. When the compressions reach the ear of the listener, they are converted to nerve impulses, which are then sent to the auditory center of the brain.

Sound and radio communications are clearly very dissimilar. Yet they also share some common features. They both propagate using wave motion

Table 1.1 *Differences Between Sound and Radio Communication*

	Sound	Electromagnetic
Wave motion	Mechanical (oscillation of molecules)	Electromagnetic (oscillation of electric and magnetic fields)
Medium	Typically air; but liquids and solids are also possible	Any matter possible but none required (i.e., it can also travel through a vacuum)
Wave direction	Longitudinal	Transverse
Speed	331.6 m/sec	300,000,000 m/sec
Frequencies	15–20,000 Hz	1–10^{24} Hz

at a finite speed and are subject to many of the same obstacles, such as interference, reflection, and refraction. Table 1.1 gives a short comparison of some of the main differences.

1.7.4 Electromagnetic radiation spectrum

The easiest and most common classification of electromagnetic radiation is according to frequency/wavelength. The delineation between the ranges may seem somewhat arbitrary. After all, our range of numbers is continuous and there is no intuitive reason to expect a mathematical function, such as a sine wave, to change behavior at different frequencies.

The reasons for the differences occur from the varied properties of matter. Each substance absorbs, reflects, and refracts different frequencies in many varied ways. For example, some waves (e.g., radio waves) are able to penetrate most solids, while others (light) can penetrate some solids (e.g., glass) but not others (e.g., brick). (See Table 1.2.)

Light

Light is the most familiar form of electromagnetic energy. It provides our visual system with information about the shape and texture of the surfaces of physical objects. Our eyes are able to translate the radiation into nerve impulses, which our brains process to give us an understanding of the material world that surrounds us.

Because of its natural occurrence and our critical dependence on it, light has been the subject of investigation throughout scientific history. Much of its behavior is still not thoroughly understood. Sometimes it behaves like a

Table 1.2 *Classification of Electromagnetic Radiation*

	Frequency (Hz)	Wavelength (m)
Radio waves	$1–10^{12}$	$10^{-4}–10^{8}$
Infrared	$10^{12}–10^{14}$	$10^{-6}–10^{-4}$
Visible light	$5–7 \times 10^{14}$	$4–7 \times 10^{-7}$
Ultraviolet	$10^{15}–10^{17}$	$10^{-9}–10^{-6}$
X-rays	$10^{17}–10^{19}$	$10^{-11}–10^{-9}$
Gamma rays	$10^{19}–10^{24}$	$10^{-16}–10^{-11}$

wave, while in other respects it appears to be transmitted as a particle (called a photon).

Light has been used for communication since prehistoric times. Indirectly it enables all forms of visual communications. Without light we would not be able to see gestures and facial expressions. But light has also been used explicitly. Torches, fires, or even lighthouses all relied on light sources that were able to extend over large distances.

In the area of high-speed data communications light is playing an ever-increasing role, since optical fibers are able to provide very high bandwidth and security.

Infrared

Infrared, also a wireless communication technique using wave propagation, refers to electromagnetic radiation in the spectrum of frequencies just below those of visible light. It shares many characteristics with the light that we are able to see, but our eyes are not "tuned" to the infrared frequencies.

There have been several practical uses of infrared that are not related to communications. It is less susceptible to dispersion through fog than is visible light. This has opened up opportunities for infrared photographs of clouded stars that would be hidden to our eyes.

The fact that infrared is not visible can be considered an advantage in an adversarial situation. For example, the military has developed infrared light beams and specialized goggles that allow soldiers to see while the enemy remains in the dark.

In the household infrared is now commonplace as a signaling technique for remotely controlled television and stereo sets. It is an attractive frequency, since it is not common and is therefore less susceptible to interference than, for example, visible light would be.

Its biggest restriction is that it must be line of sight, since it cannot pass through solids or liquids. This is an advantage in that there is much less likelihood of interference from outside. However, it does limit the range of use.

1.7.5 Radio spectrum

The focus of this book is on radio waves. They form the low subrange of the Electromagnetic Radiation (EMR) spectrum. There is no real subdivision of the radio spectrum in terms of named frequencies. However, it would be

Table 1.3 *Uses for Radio Frequencies*

	Frequency (Hz)
AM	535–1,700 KHz
FM	88–108 MHz
TV	54–88, 174–220 MHz
GPS	1,200–1,600 MHz
Cell phones	800–1,000, 1,800–2,000 MHz

misleading to believe that all radio waves behave similarly and can be treated as a homogeneous block.

Radio waves over 20 GHz (under 15 mm) may be absorbed by water vapor and are therefore not ideal for long-distance communication. Most data communications use the range between 30 MHz and 20 GHz (15 mm–10 m). The range under 30 MHz (over 10 m) would also be able to support data, but it has the added advantage that waves may reflect on the earth's ionosphere to extend the range of communication. This makes them particularly useful for broadcasting.

We will be looking at the use of radio frequencies in more detail in the next chapter. However, Table 1.3 gives you an idea of some of the varied uses for radio frequencies. This table is by no means complete. Other uses include CB radios; garage-door openers; radio-controlled cars, boats, and airplanes; cordless phones; radars; and even space communications.

1.8 Radio transmission

A typical radio communication system has two main components: a transmitter and a receiver. The transmitter generates electrical oscillations, which may be modulated (by varying either the amplitude or the frequency) to carry a signal. At the other end is the receiver, which decodes the modulated signal.

Most wireless data communication is two-way, meaning that the radios at each end include both a transmitter and a receiver. However, it does not need to be so. In cases where it is sufficient to transmit data in one direction only (e.g., for paging or Global Positioning System [GPS] satellite navigation) then one end would only need a transmitter and the other only a receiver.

1.8.1 Radio transmitter

The four main components of a transmitter are as follows:

1. Transducer
2. Oscillator
3. Modulator
4. Antenna

Transducer

A transducer converts the information to be transmitted to a varying electrical voltage. In the case of sound this would be a microphone, which converts the mechanical energy of successive air compressions into an electrical signal. Picture transmissions use a photoelectric device to capture light patters and convert them into a transmittable signal. In the case of data transmission (e.g., from a computer) there is often no need for a transducer, since the information is already in electronic form.

Oscillator

An oscillator generates a consistent and reliable frequency that is used to carry the signal. This signal is called the carrier wave. Often quartz crystals with definite natural frequencies are used, since they can produce extremely stable waveforms and frequencies.

Modulator

A modulator varies the carrier wave to project the desired signal on the transmission. If the signal is an analog waveform, the modulator can modify either the amplitude or the frequency of the carrier wave to attain this. A much simpler form of modulation is called *keying*, which involves interrupting the carrier wave. This is a binary action and, therefore, only possible for digital transmissions.

Antenna

An antenna radiates an electrical signal into space in the form of electromagnetic radiation. Every electrical current is surrounded by an electromagnetic field, which varies according to the strength of the current. The antenna is the true air interface. It is connected to the rest of the radio with two conductors, which allow electrical current to pass through them. In its characteristic shape it is a long wire that exposes an elongated electromagnetic field to the surrounding air. And this field radiates naturally in all directions.

1.8.2 Radio receiver

The four components of a receiver are very similar to those of a transmitter, as follows:

1. Antenna
2. Oscillator
3. Demodulator
4. Amplifier

The reception process is very similar to the transmission, with the obvious distinction that it operates in reverse.

Antenna

The antenna captures radio waves and converts them into an electrical signal. In this case it is the electromagnetic field, which induces electrical current in the wire. The antenna is exposed to the air, which carries the electromagnetic waves of the transmitter. These waves are transformed into an electrical wave, which the antenna passes on to the rest of the radio through two conductors.

Oscillator and demodulator

The oscillator generates electrical waves at the carrier frequency to be used as a reference wave in order to extract the signal. This is done by the demodulator, which detects modulated signals, mixes them with the oscillator signal, and restores them into their original form.

Amplifier

The received signals are usually very weak. Depending on their use, it may then be necessary to include an amplifier. For example, in the case of an FM radio, it would be necessary to amplify the signal before it is passed on to a loudspeaker, since the actual induced signal from the antenna would not be strong enough to be audible to the human ear.

Note that while the transmitter and the receiver are described here as completely separate entities, it is possible to use some of the components, such as the antenna and the oscillator, for both functions.

1.9 Wireless network architecture

1.9.1 Wireless topologies

There is no limit to the number of possible wireless topologies that could be constructed. However, most of them use one of two basic approaches: point-to-point or networked.

Point-to-point

A point-to-point configuration, also called ad hoc mode, involves a minimum of two devices equipped with radios that use the same air interface standard. It is typically a very simple scenario to implement and is intended for situations where users interact with each other in an ad hoc fashion, such as a meeting room.

All users need to do is enable their wireless interface and they are ready to share files or run simple peer-to-peer applications between them. (See Figure 1.6.)

The advantage of the scenario is its simplicity. It is a full-mesh topology, which does not require any infrastructure to route data between the end points. Its limitation is that it only provides access to the network resources of the participating devices. It isn't possible to access the Web, run remote applications, or send e-mail, for example.

Networked

A more permanent approach is needed to provide complete network access. In particular, there needs to be a link between the wireless network and the fixed public or private network.

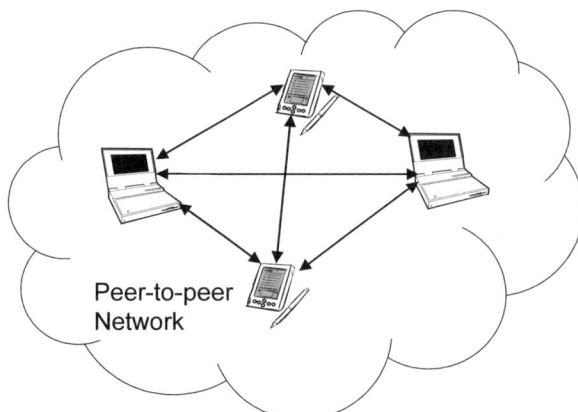

Figure 1.6 *Peer-to-peer network.*

Figure 1.7
Wireless and fixed networks.

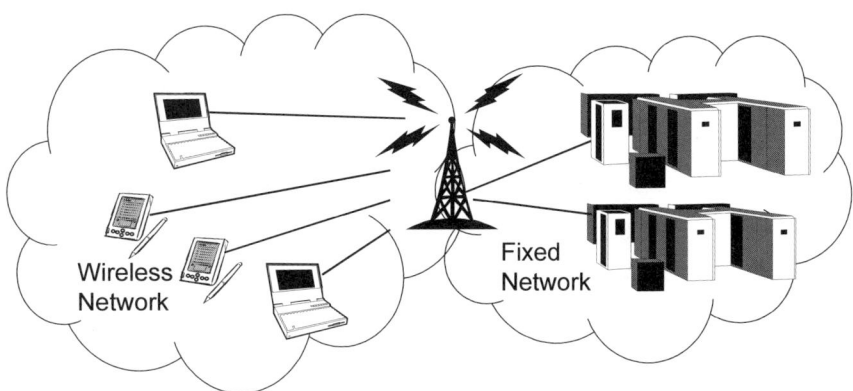

The most typical configuration, sometimes called *infrastructure mode*, involves wireless terminals, at least one bridge (usually called base station or access point) between the air and the physical network, and any number of servers hosting applications that are used by the terminals. (See Figure 1.7.)

Wireless terminals connect through the air. A transceiver will provide the connection to the fixed network. The manner in which it connects to the Internet and any specific application servers is generally through copper or optical connectors. It is worth noting, though, that there have been inroads into broadband wireless. In the future, we may see the network portion in Figure 1.7 also being implemented with wireless technology.

1.9.2 Network layers

The International Organization for Standardization (ISO) developed the most common model of a network stack in use today. It is called the Open Systems Interconnection (OSI) reference model and defines seven layered modules of a networked system. It should be possible to define all wireless connectivity in terms of this network stack. The functionality provided by a wireless system is very similar to that provided by a fixed network. (See Figure 1.8.)

However, the OSI model was designed without wireless networks and mobile devices in mind. While it can accommodate these to some degree, it would be misleading to impose an outdated structure onto some of these newer technologies. Rather than attempt to rigorously apply the OSI model, I have decided to use a less formal approach.

In Figure 1.8, I have proposed a division of some of the wireless components and ordered them into layers. We will go through each of these in the

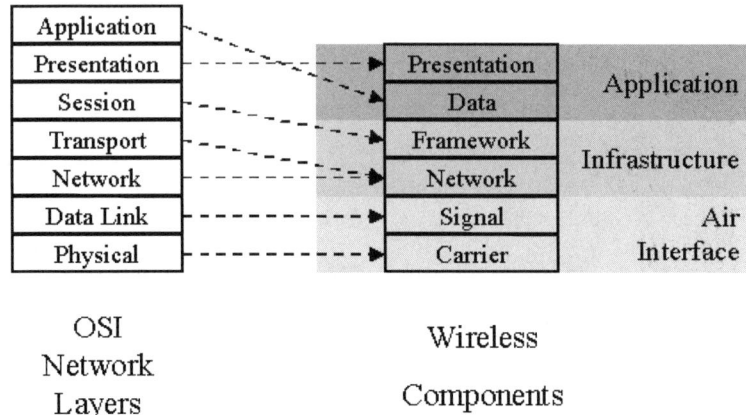

Figure 1.8 *The OSI reference model and wireless components.*

coming chapters. The arrows demonstrate that there is a vague relationship between these modules and the OSI stack. Be careful, however, not to interpret too much into these, since the correspondence is primarily conceptual.

1.10 What are the implications of wireless?

In simple terms, wireless technology provides a form of point-to-point or networked connectivity. In this sense it is very similar to the many protocols coordinating communication across copper wires or optical fibers. You may ask yourself why it should be treated any differently. Is it not simply another network?

The notable difference is that wireless transceivers do not need a physical medium of transmission. While metal cables transmit electrical current, both wireless and optical transmissions involve specific frequency ranges of electromagnetic radiation. Again, this is clearly a difference, but it is a medium fundamental characteristic or merely a consideration for physical implementation?

1.10.1 What is an air interface?

Wireless transmissions do not require any medium at all. However, they do need to traverse space to arrive at their destination. And this space it usually occupied by some kind of matter. As it turns out, this matter almost always includes air and often consists of nothing else. For ease of discussion, many people refer to air as the wireless medium. In the most literal sense, this is misleading. But it is more convenient than trying to be precise, so I will often adopt the same terminology.

There are several related implications to the choice of air as a medium. Each derives from one of the inherent characteristics of our atmosphere.

Single

There is only one instance of a given spatial expanse. It is possible to simultaneously connect a device to any number of cables, but it can only be connected to one instance of the space surrounding it.

Shared

Any other devices in the immediate vicinity also share the same medium and should be compatible with each other. In order to ensure privacy, wireless communication must include authentication and encryption.

Ready

Space is already there. While the installation of cabling is a major component of other forms of connectivity, there is no comparable effort or cost associated with making use of the air. Similarly there is no equivalent to the maintenance of wiring, since the air cannot be "broken." While atmospheric conditions may occasionally cause transmission errors, this is a situation that resolves itself without intervention.

Ubiquitous

The air has no fixed connection points. Consequently the device does not need to be stationary in order to use the connection.

1.11 Summary

Data transfer through the air is not a new phenomenon. In fact it is the most common medium for interpersonal communications and has been in use through history.

With the evolution of science many new media, including copper wires, were introduced that provided much more efficient means of communications and have been at the core of our information infrastructure for the past century.

Radiation is not a new or human-inspired technology either. It surrounds us, for example, in the form of solar energy and is clearly a critical part of our environment.

Electromagnetic radiation, specifically in the radio-frequency range, is suitable for communication because it provides a means of transmitting data through matter using an existing medium: Air. This reduces cost, since no cabling is necessary. And it can also facilitate mobility, since no fixed line tethers the users to a specific point.

Wireless networks come in two forms: They can be ad hoc networks, which connect a small set of wireless devices to each other. Or they can use a base station, which is associated with a fixed network, in order to establish a connection with the Internet, thereby enabling a much more extensive set of services and applications.

2

Air Interfaces

This chapter begins with an overview of the dilemma of how to assign ownership of spectrum over a particular area. As a consequence of multiple conflicting requirements, it is possible to divide spectrum according to both frequency and territory, and both will be discussed.

Unfortunately, it is not possible to use one common mechanism for transmitting all wireless communications. We will look at some of the needs and differentiating components of an air interface. Using these metrics, we will give an overview of some of the most common wireless standards in operation today, both for terrestrial and satellite communications.

2.1 Air ownership

The biggest problem with air is ownership. You may think the air on your property belongs to you. After all, most other things, such as buildings and trees, do. And chances are, if you have some cables or other media for communication, then those belong to you too. Or at least you have some discretion over them.

There are at least two reasons why air space is approached differently, particularly when it comes to radio transmission.

One problem with dividing along territorial lines is shown in Figure 2.1. Air transmissions do not respect artificial property boundaries. Instead they cover a range whose shape depends a lot on the obstructions it encounters but which usually approximates a circle. If you assume four neighboring lots, you can see that you cannot map the range to the land. Either it only covers a fraction of it or else it will encroach on your neighbor's space. This dilemma is fundamental to wireless and has many implications to all wireless technologies, particularly in the area of security.

Figure 2.1
Air Transmissions do not respect artificial property boundaries.

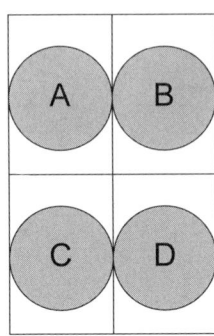
Surface not covered

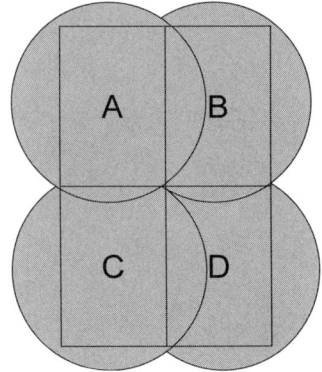
Boundaries exceeded

The second consideration isn't so much a problem as an opportunity. One of the main benefits of wireless communications is that it allows us to transmit across large distances. That is obviously one of the attractions of radio and television. But clearly it needs to cut across individual air spaces for the good of the public as a whole.

2.2 Licensing airwaves

Of course, just knowing that ownership does not come from the underlying property doesn't help answer the question of who should own the airwaves. Each country can come up with its own scheme of assigning ownership within its own territory.

In theory there could be considerable friction along common borders where transmissions interfere with the neighbor's broadcast. In practice, the responsibility typically lies with some government agency. And since all government agencies adopt very similar approaches, it has been possible to come to international agreements so that apart from the occasional jamming—for example, during the cold war—interference is not a frequent source of conflict.

There are really two approaches that can be used to segment the air over a particular geography. One is along territorial lines, and the other is through apportioning radio spectrum.

Spectrum is usually allocated by a national agency. The task is not trivial, since the amount of spectrum available is limited. Once it has been allocated it is difficult to revoke. An overly sparse allocation will waste valuable spectrum, while an overly dense allocation may result in frequent interference of adjacent applications. It is, therefore, necessary to carefully plan frequency allocation.

Once a range has been allocated for a given application, it is often still necessary to subdivide the range into bands and grant these to individual companies in order to limit interference in the range.

These bands are often licensed for a price. In an effort to ensure that the price is correct and there is no favoritism in selecting the company, bands are often auctioned and licensed to the highest bidder. During the year 2000, auctions of the Universal Mobile Telecommunications System (UMTS) licenses attracted significant attention due to the high prices they have fetched.

One very important point regarding frequency ranges is that the spectrum has not been allocated in the same way in different countries. This poses inter-operability problems, which usually lead to consumers being unable to use the same devices in different countries. Since this is clearly at odds with the mobile aspect of the wireless Internet, there have been many recent international efforts to pursue common frequency ranges.

2.2.1 Territorial division

A division of the territory is similar to the first suggestion in this chapter, but on a much larger scale. Rather than assigning ownership of the air to the property owner, the whole region is subdivided into smaller areas. Rights to the region are then conferred—for example, through an auction—to an assigned organization.

This organization is then able to decide who can transmit what over this air space. In the simple form described, this approach is rather extreme, since it gives individuals very little choice over what they can transmit and receive. Depending on their location, they would be virtually assigned to a particular operator.

2.2.2 Spectrum division

Another approach is to compartmentalize the radio frequency spectrum. Fortunately, there is very little interference between two transmissions if they use different frequencies. It is possible to take advantage of this feature to simultaneously broadcast in many different frequencies over the same area. With this system the full spectrum is subdivided into ranges, which are delegated to specific purposes and/or entities.

2.2.3 Licensing issues

In practice, licensing of the airwaves is not done on an exclusively geographical or spectral basis but uses a combination of both, along with additional stipulations.

At a minimum licensing will designate exactly who owns a given frequency range in a particular area. But, in order to ensure that there is no interference with neighboring frequencies or areas, there are usually additional limitations, such as the power of transmission, the location of transmitters, and even the techniques and purposes of the transmission.

Rather than licensing the spectrum, it is also possible to leave it unlicensed. The 2.4 GHz band, as we shall see later in the book, is probably the best known example of an unlicensed frequency. This merely means that no particular individual or organization owns the airwaves. The airwaves are open to everyone. However, this doesn't mean that they are unrestricted. In order to ensure that users of an unlicensed spectrum do not interfere with each other, specific transmission techniques are required, and maximum power levels are strictly enforced.

2.3 Ownership and billing

One approach to wireless networks is for the user (consumer, small company, or large corporation) to own and operate the network. In this case there are no usage charges. However, the owner needs to purchase the equipment and keep it operational.

Public wireless networks are owned and operated by an independent company that offers services to any registered subscribers. The consumers must pay air time, air traffic, or a fixed subscription fee. These networks typically span very large geographical areas. In some cases (particularly through roaming agreements) they may even span national boundaries.

2.4 Interface characteristics

There are a number of different radio frequency standards in operation. The primary difference is in the purpose to which they are put to use. Each has different needs and is therefore implemented differently. I have divided up some of the primary characteristics and classified them into two groups, as shown in the following chart. The first set is the physical considerations of the carrier wave. These approaches are typically more closely regulated,

2.4 Interface characteristics

since they tend to impact other users and can be cause for contention. They relate to the basic elements of the radio itself.

The other set comprises the distinctive features of air interfaces that are less related to the physics and more closely related to the data being transferred and the mechanism of encoding and optimizing these data.

There is no clear boundary between the two, but it is helpful to realize that the specifications serve at least two different purposes.

Carrier Wave	*Data Signal*
Frequency	Modulation
Transmission Power	Multiple Access
Range	Speech Encoding

For those of you who are accustomed to the layered networking models, I would describe them with the following chart, which shows the two lower layers of the stack introduced in Chapter 1.

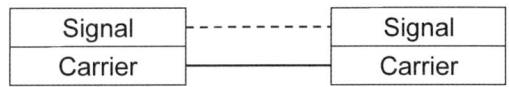

2.4.1 Carrier wave considerations

The physical considerations of licensing airwaves have already been discussed in Section 2.2. They primarily have to do with the frequency range allotted and the amount of power that can be used during transmission. There are some practical implications of both of these beyond simply defining how the radio would need to be implemented.

The simplest way to think of it is that increased power will extend the range. And a larger set of frequencies means more bandwidth over the range of transmission. But it is not always quite that simple, since each of these factors is inextricably related with the others.

Radio frequencies

Radio frequencies occupy the range of the electromagnetic radiation spectrum below infrared—that is to say, from approximately 3 KHz to 300 GHz with wavelengths from 1 mm to 100 km. This is actually a very large range and should be enough to accommodate an enormous amount of bandwidth, if it were all available.

Some parts of the spectrum are taken by some of the more commonly known uses of radio frequencies, including those shown in the following chart.

AM radio	0.535–1.7 MHz
Short-wave radio	5.9–26.1 MHz
Television stations	54–88 MHz, 174–220 MHz
FM radio	88–108 MHz

But there are many more applications, which also need their own dedicated frequencies in order to operate. Some examples include the following:

- Garage door openers, alarm systems
- Standard cordless phones
- Baby monitors
- Radio–controlled airplanes and cars
- Satellite navigations systems

Other than just needing some amount of spectrum, we need to take into account which frequencies are suitable for which applications. In particular some frequencies lend themselves to different ranges of transmission.

Range

An important characteristic of a particular wireless technology is the range in which it is intended to operate. Some frequency ranges do not lend themselves well to long distances or require line of sight in order to function. As such there is usually a maximum bound to the range at which any frequency can be used.

As mentioned previously, another critical factor is the power that is used to generate the signal. Often government regulatory agencies have placed an upper bound on the power that can be used for any given application. Although the two are not directly proportional, a simplified approximation of their relationship is that each mW corresponds to about 1 m in radius (1 W ~ 1 km, 100 mW ~ 100 m, 10 mW ~ 10 m, and 1 mW ~ 1 m).

Of course, there is typically not a thin line delineating the reach of the transmission. In practice, the signal degrades as the distance increases until it is no longer detectable. What this means is that if we are transmitting data, these data will likely incur a higher error rate the farther out we go. If we are willing to sacrifice some of our spectrum to error correction, we can obtain a higher range.

Another approach to extending the range is to use multiple base stations. This implies a much more complex infrastructure, which is able to deal with issues such as the following:

- Base station selection—how does the end device determine which base station to use when more than one is in range?
- Hand-off—how does one base station transfer the session to another when the user moves out of its range?

Depending on the technology, there may even be the possibility that the base stations are not owned by the same company and are independently controlled, thus making the task even more complex. In order to support these needs there is usually a requirement for a high-performance network infrastructure for signaling and switching.

Mobility

One of the main purposes of wireless technology is to facilitate mobility. This implies a miniaturization of the device, which means a need to reduce the battery size and therefore to minimize power consumption. (See Figure 2.2.)

The goal of reducing power consumption further implies that the generated signal will also be broadcast with less strength, which, in turn, means that the range cannot be as great.

This is one of the main reasons why there are different wireless standards. There are some applications, such as connection of a headset to a

Figure 2.2 *Facilitating mobility.*

phone, that must use very small transceivers and do not require a wide range of operation. On the other hand, fixed and automobile-mounted wireless transceivers may have no difficulty supplying power to larger devices and would thus benefit more from a greater range.

One point to consider is that not all transceivers need necessarily be the same size. It is very typical for the base stations to be larger, and transmit with more power, than the mobile devices. Since they have larger antennas, they are better able to pick up signals and can receive the weaker transmissions of the mobile devices.

Bandwidth

The bandwidth is most closely related to the amount of spectrum that is available per user. One of the easiest ways to get more data throughput is to use wider channels. But there are also other factors to consider. One of the primary obstacles to obtaining the theoretical bandwidth is the introduction of transmission errors, due either to insufficient signal strength or outside interference. (See Figure 2.3.)

When the error rates are high, a larger proportion of the transmission must deal with error detection and correction, reducing the size of the payload. Alternatively, the range of transmission can be reduced (by placing the transmitters closer together) or the transmission power can be increased. These will also serve to deliver a more reliable signal.

Beyond these physical dimensions, the amount of bandwidth is also influenced by some of the data signal components, as we will see in the next section.

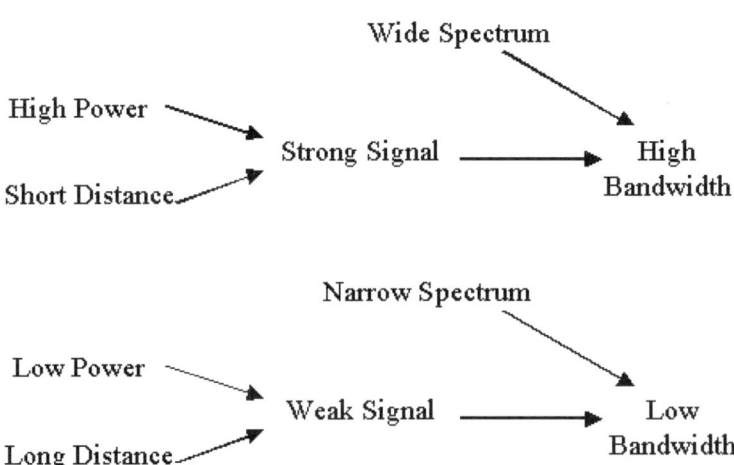

Figure 2.3
Wide and narrow bandwidths.

2.4.2 Data signal considerations

Of course, a carrier signal by itself is not of much use. It is only a means of transmitting information. How this information is imposed on the carrier frequency is different for virtually every wireless technology.

Some of the main points of differentiation include the following:

- Modulation
- Multiple access techniques
- Speech encoding
- Packet versus circuit networking

Modulation

Modulation is the application of a signal onto the bearer frequency. Until recent years most modulation was analog and was either Amplitude Modulated (AM) or Frequency Modulated (FM).

With the advent of digital data, most modulation now uses one of many forms of digital modulation called keying. While a discussion of the relative merits of each is beyond the scope of this book, each may be positioned at different points in the scale of a fundamental trade-off between power efficiency and bandwidth efficiency. In order to reduce the power needs of a modulation scheme, it is necessary to consume more bandwidth. And, vice versa, in order to optimize the bandwidth needs, a stronger signal will be required.

Keying

The three fundamental types of keying are Amplitude Shift Keying (ASK), Frequency Shift Keying (FSK), and Phase Shift Keying (PSK). They each impose binary 1s and 0s on the carrier frequency. In the case of ASK this means that the amplitude of the wave is modified. FSK varies the frequency, whereas PSK shifts the phase of the waveform. (See Figure 2.4.)

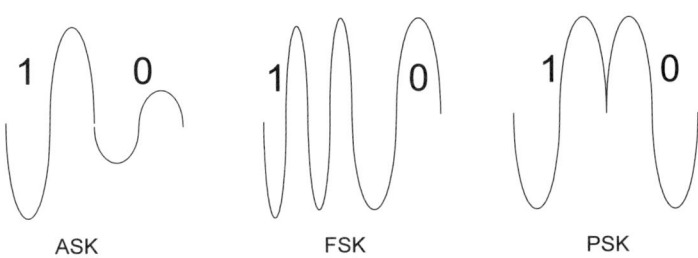

Figure 2.4 *Types of keying.*

Two of the more popular keying methods are GMSK and DQPSK. Gaussian Minimum Shift Keying (GMSK) is a type of FSK that uses continuous phase modulation so that it can avoid abrupt changes. It is used by both Global System for Mobile Communications (GSM) and Digital Enhanced Cordless Telecommunications (DECT), which we shall see later in this chapter. Differential Quadrature Phase Shift Keying (DQPSK) is used by IS-136 Time Division Multiple Access (TDMA) in the United States. As its name indicates, it is a form of PSK. Rather than only allowing two phases it defines four phases, which means that two bits can be transmitted with each phase shift.

Spread spectrum

Some of the newer wireless technologies, including wireless local area networks (LANs), employ spread spectrum modulation techniques, which use transmission bandwidth many orders of magnitude greater than required, with the result that they are very bandwidth inefficient for a single user but very efficient for multiple users. (See Figure 2.5.)

This may seem like extra work for no real gain. After all you still only have the same amount of bandwidth that you would with staying on the same frequency. You still need to modulate (or key) the individual slots, so why make the effort?

One advantage is resistance to interference. Since the transmission is not restricted to one frequency, it is unlikely to be blocked by another signal. It may lose a few units of data, but those can be recovered with good error correction.

The other main drawing point, and its reason for invention by the military, is that it is implicitly more secure. Unless someone else knows the pattern in which you are switching frequencies, it will not be easy to intercept

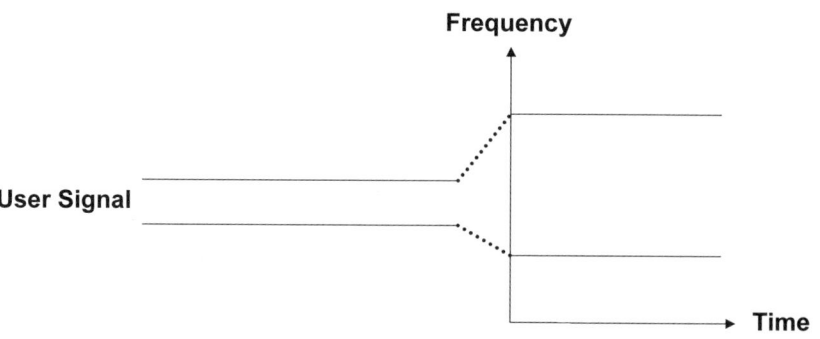

Figure 2.5
Spread spectrum modulation.

Figure 2.6
Frequency hopped spread spectrum.

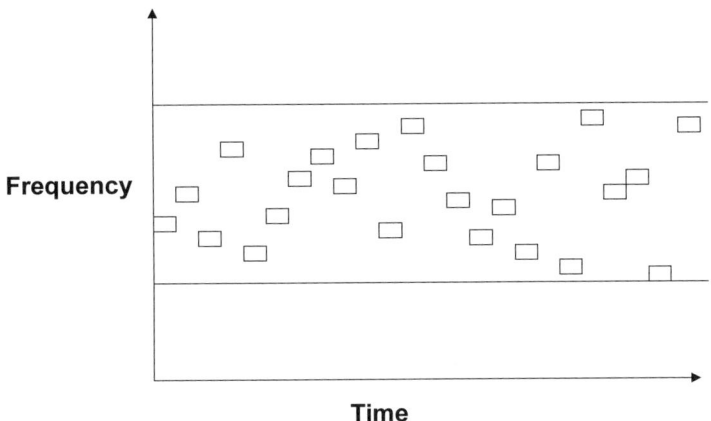

your transmission. Due to its immunity to narrowband interference it is possible to transmit at levels scarcely above the ambient noise, making it very difficult to detect any signal at all.

There are two main types of spread spectrum: DSSS and FHSS.

Frequency Hopped Spread Spectrum (FHSS) requires multiple frequencies to operate. It periodically changes the transmission frequency. The signal is then spread across the entire spectrum by dividing it into time slots. The transmission of each time slot occurs at a different frequency. (See Figure 2.6.)

Direct Sequence Spread Spectrum (DSSS) uses a binary sequence known as a chipping sequence to encode user data and spread these data over a larger frequency range. To do this, the original data signal is mixed with a second signal, which is much wider in frequency. The pattern of the second signal, the chipping sequence, is made up of a pseudo random code

Figure 2.7
Direct sequence spread spectrum.

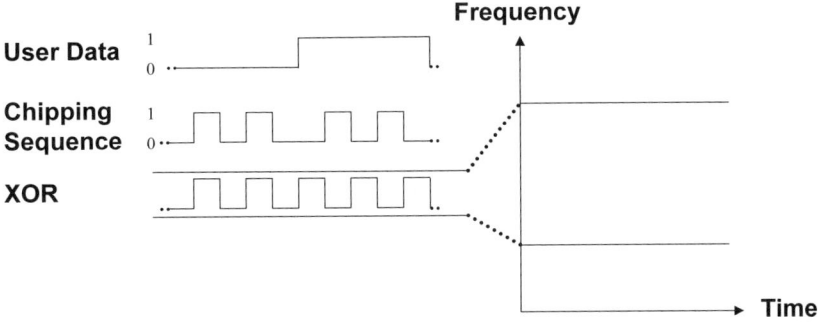

(PRC). The PRC appears as noise to systems that don't know the code. The resulting signal is as wide as the chipping sequence, but still carries the data that were contained in the original signal. By mixing the two signals together, the resulting transmission still looks like noise to any system that doesn't know the chipping sequence. (See Figure 2.7.)

While spread spectrum is becoming increasingly popular in wireless technology, the respective advantages and disadvantages of DSSS and FHSS are frequently a source of discussion and controversy.

Multiple access techniques

In order to optimize the frequency spectrum most wireless technology offers the possibility of more than one use for any given frequency range in a single geographical area.

The contention for the spectrum can be handled in one of three different ways, as follows:

1. FDMA—Frequency Division Multiple Access
2. TDMA—Time Division Multiple Access
3. CDMA—Code Division Multiple Access

FDMA divides a single band into frequencies and associates each channel with one of these frequencies. An example of FDMA is AMPS, the most common analog standard in the world today.

TDMA divides a single channel into a number of time slots. The best known TDMA standard in the world is GSM. IS-136, DECT, and Integrated Digital Enhanced Network (iDEN) also employ TDMA.

CDMA uses a spread spectrum technique as described earlier. Some benefits of the CDMA approach include the following:

- Eight to ten times more users than traditional FDMA/TDMA
- Better signal quality
- "Soft" capacity limit—additional users add more noise to the cell, thus "gracefully" degrading performance rather than blocking all new calls once a "hard" limit is reached

CDMA is the most prevalent digital standard in the United States. It was developed by Qualcomm, which holds numerous patents on the technology. Due to its superior performance over FDMA and TDMA, all mobile phone standards currently being developed are based on CDMA.

2.4 Interface characteristics

Are TDMA and CDMA standards or technologies?

The Interim Standards IS-95 and IS-136 can be a source of confusion when they use the labels TDMA and CDMA. In North America the term CDMA is not only used to designate the multiple access mechanism described previously, but, also it is a name for IS-95, which was the first standard to employ this technique. Similarly, TDMA often refers to IS-136, even though GSM and other standards also use TDMA technology.

Just to complicate matters a little bit more, IS-136 is also often called D-AMPS (Digital AMPS), since it was developed as a digital upgrade to the analog AMPS already in operation.

Speech coding

While most standards are currently focused on the transmission of digital data, one of the main wireless applications, namely telephony, has an inherently analog origin.

Voice, or any audio information, needs to be digitized before it can be digitally modulated. A comparison of the various waveform coders is beyond the scope of this book. A simplistic characterization is that higher bit rates will produce better quality (which typically varies anywhere from 1.2 to 64 Kbps). However, subjective tests indicate performance characteristics that are highly speaker dependent.

One important distinction is between vocoders, which are optimized for human voice transmissions, as opposed to other coders, which treat all audio neutrally.

Circuit versus packet data (voice versus data)

In the past, most networks have been dedicated to either voice or data transmissions. Those oriented toward data traffic did not require a constant connection but, instead, were concerned with the most efficient technique for transmitting a large amount of data in a short time. On the other hand were the phone networks, which required a very constant amount of data to be transmitted per connection. In conceptual terms, circuit switching is similar to placing a phone call. Packet switching is similar to sending data on the Internet.

Packet data networks break down a stream of data into packets and broadcast them across their network where they are received by the client devices. This is an efficient means of transmitting large amounts of data, but it does not provide the quality of service required for voice transmissions. The most common packet data networks in use today are mainly used for paging.

Most of today's mobile telephones are less efficient in data transfer than packet data networks (since they usually require a dedicated circuit that blocks an air channel even during idle time). However, they also have one advantage in that they also are able to transmit voice reliably.

Future networks will be packet-based, but they will use high-performance infrastructures, which allow them to guarantee a high quality of service. This will make it possible to effectively cater to both packet and voice transmissions.

2.5 Current interfaces

The following sections give an overview of some of the more common air interfaces that are currently deployed. The goal is to make you aware of the different standards and understand how they compare with each other. The detail specifications are beyond the scope of this book.

As indicated in Table 2.1, the simplest division of the air interfaces is according to the typical range that each covers. As with fixed networks, the boundary between them is more conceptual than explicit. For example, LAN range is conceptually unlimited in the same sense as WAN. You can set up an arbitrarily large network of access points and roam among them. The differentiating factor is only that the LAN islands are likely to be discontiguous in the short term. A cellular, wireless WAN infrastructure attempts to supply a given geographical area with comprehensive coverage.

Table 2.1 *Air Interfaces*

Network Type	Personal	Local	Metropolitan	Wide
Acronym	PAN	LAN	MAN	WAN
Direct Range	~10 m	~100 m	~30 km	unlimited, cellular
Examples	IrDA	802.11b	LMDS	GSM
	Bluetooth	HomeRF	MMDS	IS-95
		DECT		IS-136
		HiperLAN		PDC

2.6 Wireless PANs

Personal Area Networks (PANs) help connect devices usually belonging to one single individual and spanning a very small area (several meters).

2.6.1 Infrared

One of the most common short-range technologies already in use today does not use radio frequencies. There are already over 100 million electronic devices that transmit data with infrared frequencies. These include notebooks, printers, digital cameras, cell phones, and Personal Digital Assistants (PDAs).

Infrared technologies can be classified as either diffuse or direct. Diffuse infrared does not require direct line of sight. However, it is also based on visible light and therefore can only be used in the same room. Direct infrared requires direct line of sight. IrDA (from the Infrared Data Association), the standard according to which most infrared devices in the world operate, is an example of direct infrared.

While infrared is in widespread use today, it is losing ground to radio frequency devices, which are considered more reliable, are able to cover larger ranges, are able to pass through obstacles (such as walls), and support better mobility. Nonetheless, some of the protocols may continue to be used by the emerging technologies (such as IrOBEX by Bluetooth).

2.6.2 Bluetooth

Harald Blåtand (Bluetooth) was a tenth-century Danish king who united the Nordic countries and expanded the rule of the Vikings into its neighboring territories. The optimistic implication is that this technology will also unite competitors (some of which are situated in Nordic territory) that are working in personal networking.

The Bluetooth protocol started out as a replacement for office interconnection cables (such as serial, parallel, Universal Serial Bus [USB] keyboard, and mouse cables) and infrared (IR) connections between portable devices. It is designed to be both cheap and small—an entire transceiver, including antenna, should fit on a chip, and cost less than US$10. In order to reduce the battery needs, it is also a low-power technology. Its transmitters reduce power to the minimum needed to sustain a link.

The protocol uses the 2.4 GHz range and supports digitized voice in addition to data at rates of 432 Kbps, 721/56 Kbps, or 384 Kbps over a range of about 10 m. The theoretical maximum is 1 Mb/s slowed down by forward error correction.

Bluetooth uses frequency hopping spread spectrum implemented with Gaussian Frequency Shift Keying (GFSK).

2.6.3 PAN/LAN contention

There are three different short-range wireless standards in use at present. This presents a problem, since the standards use the same bandwidth (the globally available 2.4 GHz band). In other words, they interfere with each other, making it difficult to deploy more than one standard in the vicinity of each other.

In future versions, wireless LANs may move to the 5 GHz band, which will reduce the problem. However, in the short term they must learn to coexist. This can be achieved as follows:

- Physically separating the devices
- Allowing the interference and accepting the performance degradation
- Being aware of the other standards and actively minimizing the interference

2.6.4 802.15

IEEE is active in working on wireless standards for Personal Area Networks. Rather than attempting to satisfy all needs with one standard, as has been the approach with Bluetooth, the 802.15 Working Group has spawned two separate initiative for WPANs, both limited to a range of approximately 10 meters and operating in the 2.4 GHz spectrum.

802.15.3 is developing a High Rate WPAN, capable of transmitting high-resolution multimedia with sustained bandwidth of over 20 Mbps and optional modes supporting up to 66 Mbps.

802.15.4 has less ambitious bandwidth requirements (120–150 Kbps) and is instead working aggressively to produce a very low cost (less than US$1) radio specification. This standard will be appropriate where high bandwidth is not required (e.g., keyboard, mouse).

2.7 Wireless LANs

A wireless LAN can be very similar to a fixed LAN with the exception that it uses airwaves rather than cabling to connect the devices. Wireless LANs typically span a very confined area (home, office building). Instead of a socket for a cable connector, each terminal has a small transceiver. By using a wireless LAN protocol the devices are able to interact with each other. There may also be a central point where a stationary transceiver connects to a fixed line. This point effectively works as a bridge and passes network traffic to and from the mobile devices.

2.7.1 DECT

Digital Enhanced Cordless Telecommunication (DECT) was defined as a digital standard for cordless phones. Its original name was Digital European Cordless Telephony, but this was changed in order to enlarge the scope beyond telephony and the geography beyond Europe.

It is a TDMA standard using the 1,880–1,900 MHz frequencies and can support a range of approximately 50 m.

At present it is the only standard in the IMT-2000 family that is commercially available and, in fact, is currently employed by over 45 million users in 110 countries.

One of the weaknesses of DECT is that it is not stringent enough to ensure interoperability between different implementations. In order to address this there are two supplementary standards: Generic Access Profile (GAP) and MultiMedia Consortium (MMC).

2.7.2 Home RF

In order to provide connectivity between home appliances the Home Radio Frequency Working Group (HRFWG) has defined the Shared Wireless Access Protocol (SWAP). It was derived from DECT but designed to carry data (1–2 Mbps) in addition to voice. It supports up to 127 devices over a range of 40 meters.

One of its drawbacks is that it uses the 2.4 GHz-frequency also used by 802.11b and Bluetooth. Since the latter is currently receiving more attention, the future of Home RF is not clear.

At present Home RF is favored for consumer product, while 802.11 is targeted toward office use. However, many portable devices may be used both at home and in the office. Therefore, the demand to have a single standard that works in both environments will continue to increase.

2.7.3 802.11

Standard 802.11 was developed as a replacement for wired LAN technology such as Ethernet (802.3). It typically covers the area of an office building or a group of adjacent buildings.

It generally runs the same applications as a wired LAN. Its use should be transparent to the user (other than the absence of a cable). Its main advantage is its support for mobile devices. However, an important secondary consideration can be the simpler, and cheaper, installation.

Wireless LAN technology has long been available in proprietary implementations. The 802.11b standard now provides interoperability between the products of numerous vendors.

Standard 802.11b is a revision of an original 802.11 standard. It operates in the 2.4 GHz range and can accommodate short-range transmission rates of 11 Mbps (comparable to standard Ethernet—10BaseT) and reduced rates up to a range of approximately 100 m. A new 802.11a standard will be the basis for wireless LAN speeds of 50 Mbps and higher, using the 5 GHz frequency band.

Standard 802.11b subdivides its frequency band (2.4–2.483 GHz) into several channels. Not all channels are licensed in all countries, but at least some are available in all major areas.

The original 802.11 specification was able to operate using any one of the following three physical layers:

1. Direct Sequence Spread Spectrum (DSSS)
2. Frequency Hopped Spread Spectrum (FHSS)
3. Infrared (IR)

However, 802.11b has simplified the interface so that now only DSSS is supported.

802 nomenclature

The Institute of Electrical and Electronics Engineers (IEEE) is very active in some of the wireless standards. The naming of these is often grounds for

2.7 Wireless LANs

confusion. Most of the standards are developed by the IEEE 802 LAN/MAN Standards Committee. Standard 802 itself receives its name from the fact that the standardization process began in February 1980 (1980/2).

The committee operates in multiple working groups (WGs), which are identified by the second number in the standard. The currently defined WGs include the following:

802.1 Higher-Layer LAN Protocols Working Group

802.2 Logical Link Control Working Group (Inactive)

802.3 Ethernet Working Group

802.4 Token Bus Working Group (Inactive)

802.5 Token Ring Working Group

802.6 Metropolitan Area Network Working Group (Inactive)

802.7 Broadband Technical Advisory Group (TAG) (Inactive)

802.8 Fiber Optic TAG

802.9 Isochronous LAN Working Group

802.10 Security Working Group

802.11 Wireless LAN Working Group

802.12 Demand Priority Working Group

802.13 Not Used

802.14 Cable Modem Working Group

802.15 Wireless Personal Area Network (WPAN) Working Group

802.16 Broadband Wireless Access Working Group

802.17 Resilient Packet Ring Working Group

Within a working group there are often multiple standardization efforts. These projects are identified as Task Groups and are indicated by the use of a letter. For example, the 802.11 Working Group includes the following Task Groups:

802.11a High data rate extension using Orthogonal Frequency Division Multiplexing (OFDM)

802.11b High data rate extension using DSSS

802.11e Quality of service improvements for multimedia applications

802.11f Interaccess point protocol—roaming between equipment from different vendors

802.11g High rate (22 Mbps) data transmission within 2.4 GHz spectrum

802.11h Spectrum management required for 802.11a certification in Europe

802.11i Enhanced security mechanisms to address weaknesses in Wired Equivalent Privacy (WEP)

Similarly, 802.1x, which we will discuss later in the book, is a Task Group from the High-level Interface Working Group, and is called "port-based network access control."

Evolution of 802.11

The evolution of 802.11 has already begun with the transition to 802.11b virtually complete. There are several different development efforts underway both with the objective of extending the bandwidth and improving the functionality.

The increase in throughput is split into two separate initiatives. The 802.11g Working Group is developing a standard that will use the existing 2.4 GHz Industrial Scientific and Medical (ISM) band but will accommodate up to 22 Mbps. At the same time the 802.11a Working Group has a proposal to offer up to 50 Mbps using the 5-GHz Unlicensed National Information Infrastructure (UNII) band. (See Figure 2.8.)

In addition to these projects there is also work in progress on interoperability (802.11f), and quality of service (802.11e), which could be applied to

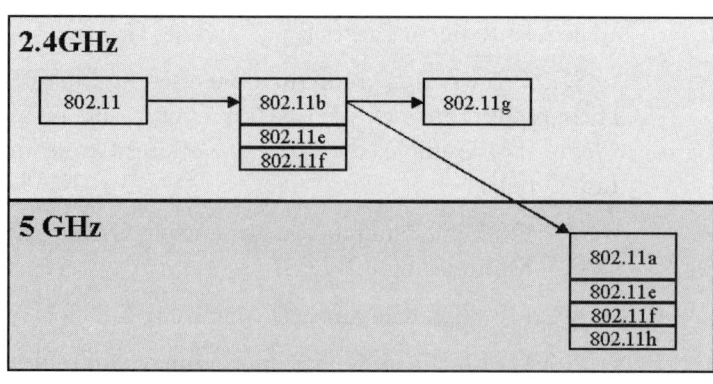

Figure 2.8 *Standard 802.11 development efforts.*

any of the approaches and is actively proposed with both 802.11b and 802.11a. Standard 802.11h is specific to 802.11a, where it offers a means to utilize a different portion of the spectrum that has been regionally allocated.

2.8 Wireless MANs

In order to extend the range of a wireless transceiver beyond the LAN the current state of technology involves power requirements and antenna sizes that prohibit mobility. Nonetheless wireless technologies for Metropolitan Area Networks (MANs) have generated a strong interest in fixed wireless techniques.

While they do not offer the mobile advantages of the other standards discussed in this book, they do deliver a means of quickly and cost-effectively providing a network infrastructure, since they do not require any physical media to be laid between the end points.

As such, fixed wireless, also called Wireless Local Loop (WLL) is becoming a competitor to Digital Subscriber Line (DSL) and cable for providing broadband to homes and small offices. The advantages are particularly appealing in rural areas, where the cost of a wireless broadband plant, on the order of US$5,000, can be minimal compared with the costs of running cable.

The two main technologies are Local Multipoint Distribution Services (LMDS) and Multichannel Multipoint Distribution System (MMDS), which both offer relatively high speeds and long range compared with other wireless protocols.

LMDS operates at a higher frequency (27.5–31.3 GHz). It provides a shorter range (around 1,500 meters) but delivers more bandwidth (up to 150 Mbps).

MMDS uses a lower frequency (2.1–2.7 GHz). It covers a wider range (up to 50 km on ideal terrain) but provides less bandwidth (up to 50 Mbps).

LMDS is very susceptible to interference from heavy rain, foliage blockage, and is restricted to line-of-sight transmissions. MMDS suffers from the same dependencies but to a lesser degree. Advances in the technology may eventually even make non-line-of-sight transmission possible.

The 802.16 Working Group is investigating commonality between LMDS and MMDS.

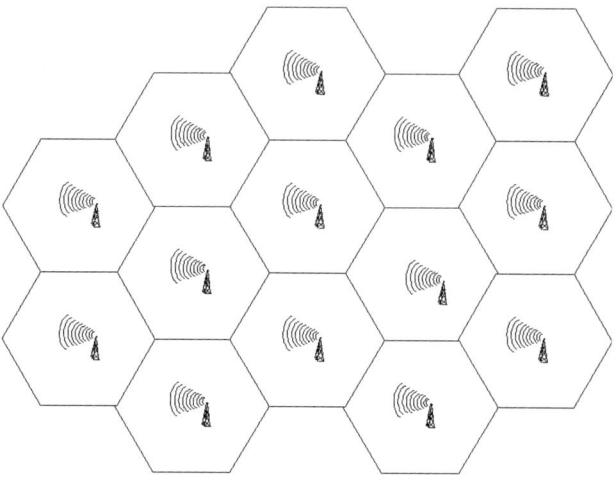

Figure 2.9
A cell is represented as a hexagon.

2.9 Wireless WANs

2.9.1 Cellular networks

In order to extend the range beyond the reach of a single transceiver, it is possible to install networks of base stations. Each of these is responsible for the coverage in its adjacent area (called its *cell*).

Given its geometric properties, the shape of a cell is most conveniently represented as a hexagon.[1] As shown in Figure 2.9, it is possible to deploy a set of base stations that provides complete coverage of any given terrain.

Most cellular networks currently fall into one of two distinct categories, as follows:

- Mobile phone networks
  - Carry primarily voice
  - Typically circuit switched
  - More prevalent in Europe and Asia
- Packet data networks
  - Carry primarily data (often for pagers)
  - Typically packet switched
  - More common in North America

---

1. See *Wireless Communications: Principles & Practice* by Theodore S. Rappaport for more information.

2.9 Wireless WANs

As we shall see, the developing standards will merge the functionality of both of these networks.

Mobile telephony

There is a wide range of standards for mobile telephones throughout the world. Analog phone systems such as Advanced Mobile Phone Service (AMPS), Total Access Communications System (TACS), and Nordic Mobile Telephone (NMT) still support millions of users. However, these numbers are quickly decreasing in favor of digital standards. (See Table 2.2.)

GSM—Global System for Mobile Communications

GSM is the dominant world standard for digital mobile telephony covering approximately two-thirds of the market. (According to the EMC World Cellular Database the total number of subscribers in June 2000 were [in millions] 494.6, broken down as 331.5 [GSM], 67.1 [CDMA], 48.2 [PDC], and 47.8 [IS-136].) While Europe, Africa, and large parts of Asia almost exclusively use GSM, it is also a primary standard in almost every area in the world.

It is a TDMA standard operating at 900 MHz, 1,800 MHz, and 1,900 MHz, depending on the country.

IS-95—cdmaOne

cdmaOne (often simply called CDMA) is the dominant standard in North America and is also used extensively in South America, Korea, and soon in Australia.

Table 2.2 *Standards for Mobile Telephones*

	Frequency	Multiple Access	Geographical Coverage
GSM	900, 1,800, 1,900 (North America)	TDMA	Worldwide (North America: only 1,900 MHz)
IS-95	900, 1,800	CDMA	North and South America, Korea, Australia
IS-136	900, 1,800	TDMA	North and South America
PDC	800, 1,500	TDMA	Japan
iDEN		TDMA	North and South America

IS-136—TDMA

IS-136 (often simply called TDMA) is the dominant standard in South America and also a primary standard in North America. It is promoted by the Universal Wireless Communications Consortium (UWCC).

iDEN

iDEN is a proprietary technology developed by Motorola and used by Nextel in North and South America. It is also based on TDMA technology. Nextel offers handsets that are also capable of GSM in order to provide worldwide roaming.

PDC—Personal Digital Cellular

PDC is a TDMA system developed by NTT DoCoMo for the Japanese market. It operates in the 800- and 1,500-MHz bands and is the primary digital standard used in Japan.

Low-speed packet data networks

Mobile data networking is not a new technology. In North America packet data networks have been in use for many years. These have often been built in parallel to the mobile phone networks, particularly for paging.

Some of the better-known mobile data networks include the following:

- Cellular Digital Packet Data (CDPD)
- Mobitex
- ReFlex
- DataTAC

They provide data transfer rates of up to 20 Kbps, which is acceptable for some applications such as vehicle location and remote monitoring. These networks have also been employed more recently for other applications, such as e-mail and Web browsing, where the data rates are only marginally sufficient.

As packet-switched data service is deployed on current and next-generation mobile telephony networks, these technologies will become obsolete.

Emerging standards

Although the current standards are not perfect in terms of voice quality, the major thrust of the emerging standards is not an improvement in the voice transmissions, but rather an increase in the data rates.

Most current standards have been voice oriented and have provided the ability to transfer data as a layer on top of the voice channel. The connections have been circuit switched and have not been tailored to data traffic.

The newer standards involve a migration from circuit-switched to packet-switched data, offering the following two important advantages:

1. Immediacy—While circuits typically take a long time to set up, packets can be sent instantaneously.

2. Spectrum efficiency—Packet switching means that radio resources are allocated only when users are actually sending or receiving data.

Newer standards, which focus on high-speed packet-switched data, are referred to as third generation (3G) mobile phone systems. This is in contrast the first generation analog and current second generation (circuit-switched) digital phone systems.

2.5G

The first step in the evolution to 3G is often called 2.5G. While there is fairly common agreement on the meaning of 3G, there are differences in opinion regarding what constitutes 2.5G. It often denotes an increase in bandwidth to approximately 100 Kbps, much higher than today's networks but far short of the expectations for 3G. At the same time it is also an overlay of both circuit (voice) and packet (data) networks. (See Table 2.3.)

GPRS

Many GSM operators are already implementing General Packet Radio Service (GPRS). It does not interfere with the GSM system in operation and does not require replacement of the base stations. As such it is a relatively low-cost add-on for current GSM providers.

Table 2.3 *The Evolution to 3G Mobile Phone Systems*

	1G	2G	2.5G	3G
	Analog	Digital	Digital	Digital
Data Type	None	Circuit only	Circuit and packet	Packet only
Data Speed		~10 Kbps	~100 Kbps	~1 Mbps
Status	Obsolete	In operation	Being deployed	In testing

Most of its challenges are not technical but related to billing and roaming agreements, since all GSM charges to date have been based on connection time rather than data transfer.

In addition to GSM, the IS-136 Time Division Multiple Access (TDMA) standard, popular in North and South America, could also support GPRS. However, it is more likely that the IS-136 operators will move directly to EDGE.

EDGE

Enhanced Data Rates for GSM Evolution (EDGE) is another high-speed mobile data standard. It allows data transmission speeds of up to 384 Kbps to be achieved when all eight time slots are used.

It offers the same advantages as GPRS, but it utilizes a new modulation scheme, which allows a much higher bit rate across the air interface. Unfortunately EDGE is not as trivial as GPRS for GSM providers to install. It requires that all base stations be replaced, which has prompted many to consider bypassing EDGE and moving directly from GPRS to UMTS.

3G

The third generation of mobile phones is expected to enable a wide variety of high-bandwidth applications and may fundamentally change the way wireless networking is approached.

UMTS

UMTS supports two different air interfaces. While both of them use CDMA, wideband CDMA (W-CDMA) will be used for the cellular wide area coverage and high mobility service, and time-division CDMA (TD-CDMA) will be used for low mobility, local in-building services, asymmetric data transmission, and typical office applications.

The evolution from today's 2G networks to the 3G networks of tomorrow is not as straightforward as one might expect. In particular there are at least two end points. While GSM, IS-136, and PDC operators have all largely endorsed the UMTS standard and expect to migrate to it, Qualcomm has developed a similar standard called cdma2000, which may also be able to draw a large following, especially from existing IS-95 carriers.

Figure 2.10 depicts some of the migration paths that can be expected over the coming years. However, it is important to realize that any operator can choose to implement any standard and need not follow these paths. In

2.9 Wireless WANs

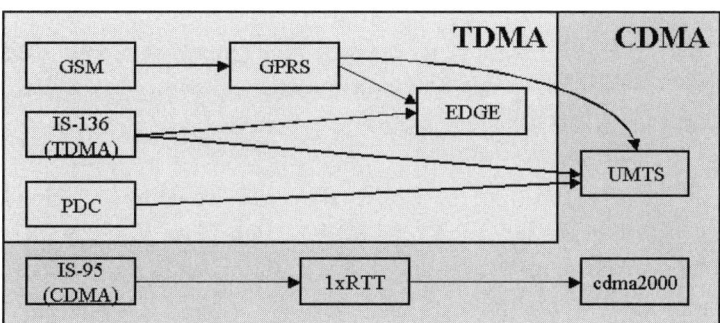

Figure 2.10
Future migration paths.

Korea, for example, the IS-95 operators have expressed intentions to migrate to UMTS.

GSM evolution

The evolution of GSM involves both an improvement of circuit-switched data (currently limited to 14.4 Kbps) and a series of packet-data technologies, as follows:

- HSCSD (high-speed circuit-switched data) adjoins multiple circuits in order to allot more bandwidth for data. With four circuits at 14.4 Kbps this produces an effective throughput of 56 Kbps.

- GPRS is a relatively inexpensive overlay to the existing GSM network and is currently being deployed by many GSM carriers with expected widespread production by 2002.

- EDGE is not an easy upgrade for GSM networks. Since it uses a different modulation technique, it will require that most of the infrastructure be replaced. Consequently most GSM operators expect to migrate directly from GPRS to UMTS. However, EDGE may well represent a viable lower-cost alternative for operators who were unable to obtain a UMTS license.

- UMTS is currently only in test phase by GSM carriers. It will not be fully deployed for several years after it is in operation in Japan.

IS-136 evolution

Despite the name of EDGE in particular this evolution is not restricted to the GSM standard. According to the current plans most TDMA carriers will also migrate to EDGE and/or UMTS.

PDC evolution

The Japanese PDC operators are already piloting UMTS. They expect to be the first to deploy it commercially, possibly as early as 2002.

IS-95 evolution

As mentioned previously the natural upgrade path for IS-95 is to cdma2000 (also called 1xEV or HDR—High Data Rate), which should offer 2 Mps. In the interim, 1xRTT is a 2.5G network, which will provide data rates up to 150 Kbps.

The danger of bandwidth projections

Most of the 2.5G and 3G standards are publicized with impressive figures regarding the maximum bandwidth available. It is important to realize that these are unlikely to represent the typical bandwidth that can actually be obtained. There are several factors that can influence the actual data rate that is possible. These include the following:

- Movement of the transceiver—If the users are stationary, the bandwidth will be higher than if they are walking and much higher than if they are traveling at high speed in a car or train.

- Position in the cell—If the user is situated in an optimal location in the cell, with direct line of sight, no interference, and not too far from the base station, the transmission will be clearer and require less error correction than if the user is at the edge of the cell in a basement filled with electronic equipment.

- Density of users—The more users that are simultaneously transmitting within a given cell, the less aggregate bandwidth is available per user.

Table 2.4 *Variations in Bandwidth*

	1	2	3	4	5	6	7	8
CS1	9.05	18.10	27.15	36.20	45.25	54.30	63.35	72.40
CS2	13.40	26.60	40.20	53.60	67.00	80.40	93.80	107.20
CS3	15.60	31.20	46.80	62.40	78.00	93.60	109.20	124.80
CS4	21.40	42.80	64.20	85.60	107.00	128.40	149.80	171.20

The maximum figures quoted typically assume a single stationary person, optimally situated in the cell. In practice only a fraction of this bandwidth may actually be available.

The variation in possible bandwidth can be seen in Table 2.4, illustrated with GPRS, which allows from one to eight time slots to be used. The amount of data that each slot can accommodate depends on the coding scheme in use. If the signal is strong and requires little error correction, then up to 21.4 Kbps are possible per time slot with CS4. On the other hand, in poor conditions only 9.05 Kbps would be transmitted per time slot.

Current implementation is restricted to only using coding schemes 1 and 2 with up to four time slots for a maximum bandwidth of 53.6 Kbps. Eventually it should be possible to attain bandwidth of up to 171.2 Kbps. However, if there are other users in the cell, then some of the time slots may already be occupied. And if the connection is not good, it may be necessary to fall back to a less-efficient coding scheme.

2.10 Satellite communications

The idea of satellite communications comes from the extended range that a transceiver has if placed at a high altitude. It is the reason that you often see radio station antennas on the top of mountains, on high buildings, or even on blimps. The higher elevation extends the horizon and opens the possibility of covering a larger territory.

A satellite is a manmade moon. It is an object placed into the orbit of some celestial body—in our case, typically the earth. While we are primarily interested in communications, this is not the only purpose of satellites. They are also used for weather (meteorological) observation and military intelligence, as well as scientific analysis of space. (See Figure 2.11.)

The primary classification of different satellites is the orbit they use. There are three different components of the orbit: altitude, shape (circular or elliptical), and path (on the earth's surface).

There are three main categories, as follows:

1. GEOs (Geostationary Equatorial Orbit)—orbit the earth with a frequency of exactly one day, so that they are synchronized with the rotation of the earth and appear stationary over a single point on the earth's surface. In order to achieve this frequency they must orbit at a very specific altitude, which is approximately 36,000 km over the earth's surface.

2. LEOs (Low Earth Orbit)—orbit the earth at an altitude of 2,000 km or less. The low altitude means less interference in transmissions to the earth's surface, less power required to transmit, and less latency in transmission.

3. MEOs (Medium Earth Orbit)—orbit the earth at an altitude of about 10,000 km and balance the benefits and problems of LEOs and GEOs.

Satellite navigation

GPS provides very accurate positioning and velocity information. It was formerly known as the Navstar Global Positioning System. It is owned and operated by the U.S. Department of Defense (DoD), but now caters to a broad spectrum of users. The system uses 24 MEO satellites spread around the world in six planes.

GPS works by computing the difference between the time that a signal is sent and the time it is received. The send time is coded in the signal broadcast by the satellite so that the receiver can calculate the difference. The receiver then calculates the distance to the satellite. The signal also contains information needed to determine the location of the satellite. By combining the distance with the location of the satellite it is able to determine its location in one dimension. By adding the same information from two

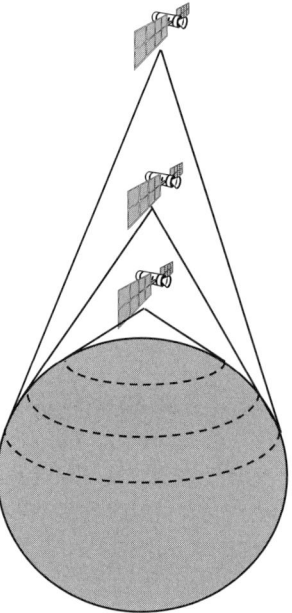

Figure 2.11
A satellite.

additional satellites it is then able to calculate its own three-dimensional position.

The most common form of GPS is the standard positioning service, which provides a position that is accurate to about 100 m. Authorized users may use a more precise form, which can narrow down the position to about 20 m. But in order to expand beyond that it is necessary to complement GPS with fixed ground stations. Enhanced techniques such as Differential GPS can provide accuracy to about 3 m and there have been tests that indicate position of within 1 cm is possible.

Satellite phones

The objective of satellite phones is not to compete with terrestrial systems, but to complement them by offering service in areas without terrestrial coverage. This means a need for dual-mode handsets, which automatically will first attempt to connect to a (cheaper and more reliable) terrestrial base station, but can go over to a satellite connection when the user is in a remote area (e.g., at sea or in the desert).

Some of the principal systems in development include the following:

- Globalstar. CDMA, 48 satellites, covering most of the world. Uses dual-mode CDMA and GSM phones offering data rates of 9,600 bps.
- ICO Global Communications. TDMA, 12 MEO satellites. 2,400 bps. Based on GSM/GPRS. Originally begun by Inmarsat.
- Iridium. TDMA, 66 satellites. 2,400 bps.
- Orbcomm. 28 satellites. Two-way low-bandwidth messaging service available today. It does not offer voice.
- Teledesic is scheduled to go into operation about 2004. It uses high-bandwidth (64 Mbps for most users) LEO satellites. 19 GHz down and 29 GHz up. Will operate with 288 satellites. Uses TDMA.
  The financial viability of these systems is questionable. Many of them have been very close to bankruptcy but have since received renewed interest from some very prominent investors. Whether this will be enough to bring them to profitability remains to be seen.

2.11 Summary

It is difficult to assign ownership to radio transmissions fairly. Airwaves do not have a clearly defined range that can be tied to property boundaries.

There is also a legitimate need to broadcast information across public and private boundaries in order to maximize its usefulness. Throughout the world, government agencies have been chartered with the task of independently licensing spectrum in a manner that optimizes the public benefit. Unfortunately, the different national agencies have not always approached the problem in the same way, which has made international interoperability very difficult.

Wireless applications themselves have many different requirements in terms of bandwidth, range, and mobility. It is therefore not practical to design a single transmission standard to fulfill all needs. Instead, there is an ever-growing number of techniques, which are differentiated according to such characteristics as transmission frequency, amount of spectrum used, or radiated power permitted. They also often use different means of sharing the carrier frequency and modulating the data signal.

One of the most common ways of categorizing all these standards is according to the range they are able to cover. Wireless PANs, such as Infrared or Bluetooth, cover very confined "personal" spaces. Wiress LANs, such as 802.11, HomeRF, and DECT, are able to cover homes and small offices. Wireless MANs, such as LMDS and MMDS, can cover entire cities. Wireless WANs include both mobile telephone systems, such as GSM, IS-95, and IS-136, and Mobile data networks, such as CDPD and GPRS. They typically use cellular infrastructure and roaming to cover large parts of the world. Beyond these terrestrial standards, the field of satellite transmissions may also provide a basis for future point-to-point communications.

Bibliography and related Web sites

Wireless technology

Rappaport, T. *Wireless Communications*. Prentice Hall, 1996.

Schiller, J. *Mobile Communications*. Addison-Wesley, 2000.

Mehrotra, A. *GSM System Engineering*. Artech House, 1997.

Air interfaces

GSM Association: http://www.gsmworld.com/

CDMA Development Group: http://www.cdg.org/

Bluetooth: http://www.bluetooth.com/

IEEE 802 LAN/MAN Standards Committee: http://www.ieee802.org/

Universal Wireless Communications Consortium: http://www.uwcc.org/

Mobile GPRS: http://www.mobileGPRS.com/

EDGE: http://www.mobilegprs.com/edge.htm

UMTS Forum: http://www.umts-forum.org/

Satellite systems

Globalstar: http://www.globalstar.com

ICO: http://www.ico.com

IRIDIUM: http://www.iridium.com

Teledesic: http://www.teledesic.com

3

Mobile Devices

The ultimate goal of most wireless technology is to support increased mobility of the work force. What this necessarily implies is that the end users must carry with them some kind of portable terminal from which they can access the unwired network. This is the common requirement that has driven the emerging market of mobile devices.

However, the uniformity stops there. Form factors, cost considerations, and the diversity of users' needs and preferences have given rise to a wide range of gadgets, making the selection, implementation, and support of mobile platforms a huge challenge to the enterprise.

This chapter begins with an overview of the different types of mobile devices on the market. This includes phones, handheld devices, and a variety of consumer appliances that are increasingly being equipped with wireless technology. We will cover the differences between these device types and mention some of the main platforms that are currently available.

The discussion will then focus on the types of applications that are relevant to a mobile environment, including personal information management and entertainment. It will elaborate on some of the issues of mobility, including the need to cater to many different machine interfaces and the requirement of synchronizing replicated data.

In a wireless environment, the paradigm for user interaction will increasingly become multi-modal. One of the machine interfaces that is of particular importance in a mobile environment is speech processing, including both automated speech recognition and text-to-speech rendering.

Mobile devices incorporate a whole new set of challenges in terms of management and support. We will look at some of the issues facing IT departments as they prepare for a wide variety of new access devices.

3.1 Device types

The primary trade-off in device types is between mobility and functionality. The larger the terminal, the more cumbersome it is to carry or wear it. Smaller gadgets may be extremely convenient but don't always have the processing power or storage needed for complex applications. We continuously see further miniaturization of all the components. Nonetheless, at any given stage of technology, there are simple physical limits reflecting the need for heat dissipation and a power source.

As shown in very general terms in Figure 3.1 Mobile devices, there is a simple trade-off between the mobility of a device and its functionality (or ease of use). It is thus largely a subjective question of user preference regarding which device, or set of devices, is selected. There is almost a continuum of devices from the headset and watch over the mobile phone, the emerging smart phone and the handheld PDA, to the laptop and eventually the desktop. Each is more powerful but less mobile than the previous. But there are also a number of special appliances that do not fit neatly into the spectrum. Car or boat phones and computers, as well as an increasing range of embedded systems for the consumer market, often have very specialized purposes and may or may not be constrained by size.

Other than size, we can also distinguish between the human-machine interfaces (HMI) of the terminals. Stationary computers very consistently

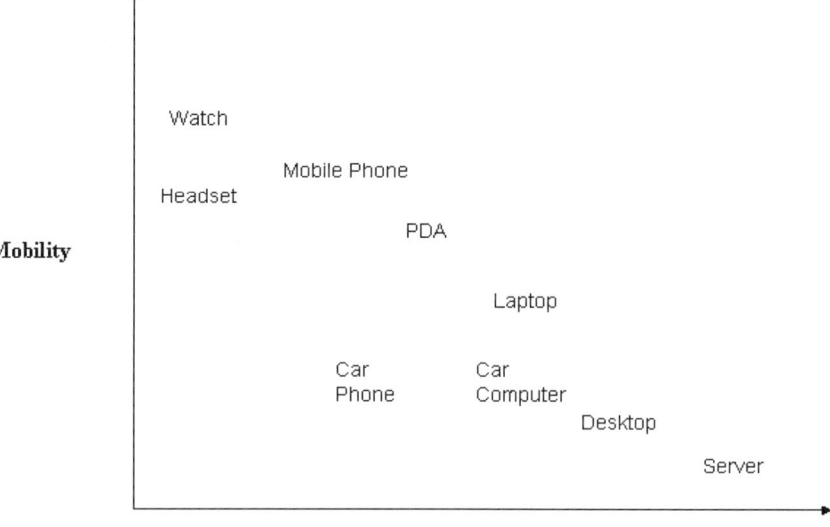

Figure 3.1
Mobile devices.

offer a keyboard and pointing device for input and a VGA screen as the primary means of output. Clearly many also provide additional interfaces via sound cards and joy sticks, but they are invariably the basic means of interacting with the user. The consistency is an advantage in a fixed environment, where the size of the interfaces is not an issue.

What is the standard Internet device?

It may seem as if there is a large amount of variation in today's desktops. They run with different speed CPUs, have different amounts of RAM, and a wide amount of disparity in hard disks. And yet the actual human-machine interface (HMI) is fairly consistent.

Almost all PCs offer a VGA screen or better. VGA is an acronym for Video Graphics Array, an indication of the resolution of the video signal being output by a personal computer. It consists of 640 vertical lines × 480 horizontal lines. Other measurements include SVGA (800 × 600), XGA (1,024 × 768), SXGA (1,280 × 1,024), and UXGA (1,600 × 1,200).

For input an application can rely on there being some kind of pointing device, usually a mouse but sometimes a trackball. A typical keyboard is equipped with all the typewriter keys but also usually with a numeric keypad as well as general function keys and special Windows keys such as Alt and Ctl.

It is also the norm for new PCs to offer some kind of sound output. Home users often improve it through a dedicated sound card and speakers, but many PCs also ship with limited embedded sound capability.

Sound input, for example through a microphone, is less common but, particularly as speech processing functions improve, we can expect it to increasingly play a role in data input.

However, pervasive appliances would lose their mobility if they used I/O mechanisms an order of magnitude larger than the mobile unit. In order to constrain their size, designers have experimented with many new and creative means of allowing interaction with humans. Some of these are listed in Table 3.1.

Phones often use a small keypad with overloaded keys, which accommodate the full alphabet either through multiple presses or through built-in dictionary comparisons (Tegic T9). PDAs usually supply a stylus, a pen-like pointing device that is used with a software keyboard and can also be used in conjunction with handwriting recognition systems.

Both phones and PDAs are increasingly utilizing speech recognition as a means of allowing user input without impacting the form factor of the

Table 3.1 *Input and Output Mechanism Features*

Input Mechanisms		Output Mechanisms	
Speech	Automated speech recognition	Screen	Embedded
Stylus	Tap selection and dragging	Monitor	External
	Handwriting recognition	Sound	Internal speakers
Keypad	Alphabet overloaded on 12 keys		External speakers
Keyboard	Internal (e.g., Thumbtouch)	Speech	Text to speech using sound interface
	External		
Mouse	Internal		
	External		
Joystick			

device. Although very attractive in ease of use, it is not suitable for all environments (e.g., airplanes, libraries), and is best used in conjunction with other interfaces.

Although there is an ever-growing diversity in handheld devices, it is possible to classify them into broad categories based on their basic shape. There is the classic mobile phone shape, which offers only a keypad and small screen. The main functional changes in the phone over the past few years are the increasing size of the display (to accommodate data, such as WAP) and the addition of supplementary function keys, which can help to navigate menus and enter shortcuts. In addition to those there have been many cosmetic and ergonomic improvements, including the reduced size of the phone, the use of color displays, and the wide range of personal choice in everything from phone covers to ringing tones.

The phone was obviously designed with voice communication as its primary purpose. If we look at devices that take a stronger emphasis on data, we can see two different approaches. There is the clamshell (the Nokia Communicator), which can be folded open and offers a small keyboard as well as an enlarged screen, usually in landscape format (much wider than it is high). Mobile phone manufacturers that want to add data communications to their phones typically adopt this approach. (See Figure 3.2.)

The second approach is the tablet PC. This type of device, which includes palm and pocket PCs, uses a (relatively) large screen, which is

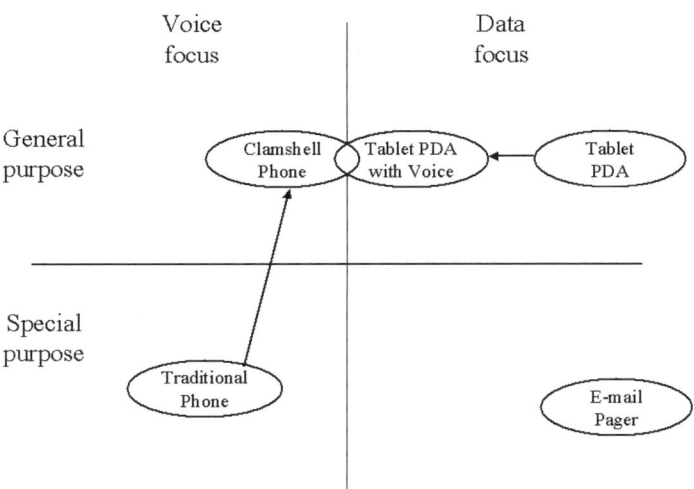

Figure 3.2
Telephony devices.

displayed in portrait format (higher than it is wide). Rather than with a keyboard, data entry is often accomplished with a stylus.

In addition to these two devices, which are slowly converging as they both support voice and data, there is also another niche product that has seen significant success in the North American market. Dedicated e-mail pagers, such as the Motorola T900 and the RIM Blackberry, are based only on data networks and are tailored for one specific application: e-mail. In order to facilitate data entry they provide miniature keyboards, which would be cumbersome for long messages but are very adequate for reading and writing short messages.

3.2 Platforms

Other than hardware characteristics, the most important classification of mobile devices is according to the operating system they employ. The list is almost endless, particularly as we extend into the closely related embedded market, including OS-9, from Microware; JavaOS and Java 2 Micro edition, from Sun; FlexOS, from Motorola; pSOS, from Integrated Systems; and Tornado VxWorks, from Wind River systems.

3.2.1 Embedded and real-time operating systems

There is a fundamental difference between a general-purpose platform and an embedded system. The latter is typically geared toward a very specific purpose and must be able to service it in real time. In order to respond with minimal latency this type of platform needs a large set of priority levels and should permit nested interrupts.

The focus of this book is on general-purpose devices; however, this does not prevent many of the concepts from also being applied to the embedded market.

The devices of the mobile appliances are driven by code similar to that of a computer operating system. In theory this system code can be proprietary, particularly since most applications will be dedicated to specific device characteristics.

However, by opening up the operating system there is the potential to expand the number of applications running on a particular appliance. These applications do not necessarily need to relate to telephony or the wireless technology.

3.2.2 PDAs

The three biggest players for strictly mobile PDA-like systems are the following:

- PalmOS—developed by Palm
- Windows CE (and Windows NT Embedded)—developed by Microsoft
- EPOC—developed by Psion and adopted by the Symbian initiative

Windows CE was designed to support a broad range of client devices combining the familiarity of Windows with real-time processing support. Some examples of Windows CE-based devices that Microsoft has defined include the Handheld PC, Pocket PC, WebTV, and Smart Phone.

What are handheld PCs and pocket PCs, and how do they relate to Windows CE?

The terminology is somewhat confusing and is often used incorrectly. pocket PCs and handheld PCs refer to specific types of mobile devices, while Windows CE is the operating system that runs on all handheld PCs and pocket PCs.

Both pocket PCs and handheld PCs contain a fixed package of integrated applications, connectivity options, and APIs for developers. The main distinction between the two is the screen size. A pocket PC has a quarter VGA (320 × 240) screen whereas a handheld PC may have half VGA (640 × 240) or full-size (640 × 480 or 800 × 600) screens with or without an integrated keyboard.

PalmOS is important because it has the largest installed base of any mobile operating system, having captured almost three-quarters of the PDA market. In addition to Palm, itself, Handspring also uses PalmOS on its Visor platform.

EPOC was originally developed for PDAs by Psion. However, it has since been adopted as the platform of choice by the major phone manufacturers. As such it is likely that it will increasingly be adopted for smart phones and PDAs developed by Nokia, Ericsson, Motorola, Siemens, and Matsushita (Panasonic).

In addition to these primary open platforms there are also a few other technologies that may surface as major players in the future. These include the following:

- Microsoft Mobile Explorer (MME) can be integrated with any real-time operating system and will work with any air interface. Its primary functionality is as a dual-mode microbrowser that can render both WML and HTML content.

- Java 2 Micro edition (J2ME) is an end-to-end Java technology aimed at the consumer and embedded market. It provides a highly optimized Java run-time environment targeting consumer products, including pagers, cellular phones, screen phones, and car navigation systems.

- Linux, which originated as a desktop and server operating system, is increasingly appearing on mobile devices. There is a version of Linux that will run on Compaq's iPAQ PocketPC. Another version, Mobile Linux, was developed by Transmeta for its Crusoe processor targeted at the embedded market.

3.2.3 Phones

While most of the phone manufacturers have endorsed the Symbian initiative, the bulk of the phones on the market still run proprietary software. As such they are not extensible and not easily upgraded with new versions.

There are substantial differences in the user interfaces, partially due to the variance in input and output facilities (e.g., the number and function of the keys). They also run different applications and simply have different command hierarchies based on the preferences of the designers. There is very little personal choice in structuring the interface, although many do offer shortcut keys that can be assigned by the user.

While the original phones were used exclusively for making calls, many now offer some primitive applications, such as personal information managers and games. Most also now have some mechanism for sending short messages and increasingly they sport a WAP browser.

Data access

In order to support WAP, or any other data communication, they need an built-in modem and the subscription must be enabled by the operator for data support. They then also need a PPP stack on the phone that can relay IP traffic from an application (or connected PC) to the dial-in server.

Another source of difference between WAP phones is the browser. Nokia and Ericsson have developed their own browsers, whereas most of the other phone vendors (e.g., Siemens, Motorola) use an Openwave browser. The difference in browser can be significant in terms of interoperability (similar to the challenge of making Web pages readable both by Internet Explorer and Netscape Navigator), since not all WAP/WML pages will display equally well on all browsers.

A fully configurable WAP phone will allow specification of the IP address of the WAP gateway and possibly also the port numbers for both TCP and UDP traffic. Additionally, if it uses circuit-switched data (a dial-in line rather than a packet data network), then it will be able to configure the phone number of the connection as well as the user's credentials.

It is worth noting that while the PPP stack is standard there is some variance in support of the authentication algorithms. In particular, Microsoft RAS servers are usually configured to require MS-CHAP, which is not supported on all phones.

3.3 Applications

When it comes to applications there is an obvious dichotomy in the current set of mobile devices. There is one group of devices—namely, the phones—that has a primary focus on communication—specifically, voice communication. The PDAs have historically had very weak networking capability but instead have offered a greater portfolio of applications.

3.3.1 Personal information management

The initial interactive applications of all the devices have been clustered around what is known as Personal Information Managers or PIMs. This information centers on the information needs of an individual. In actuality this means contact information (names, addresses, and phone numbers) and time management (appointments, meetings, and action items).

As more applications have become available on the machines, the meaning of personal information has evolved. Browser cookies are able to store virtually every preference, from travel arrangements to volume settings to favorite MP3 songs, and applications often cache recently used data such as Web pages.

As devices have become mobile, the needs have clearly also changed. Location awareness has become important. Users want location-specific news, weather, and traffic information. But there are also subtler needs, such as calendars, which are able to adjust appointments according to time zone.

Personal information is not only useful for the owner, in many cases there is value in being able to publish personal information. An example of this would be presence information along the lines of the status that is publicly visible in an instant messaging system.

Personal versus collaborative versus public information

The concept of publishing (See Figure 3.3) brings up another development in office applications. In the past most software was geared to helping indi-

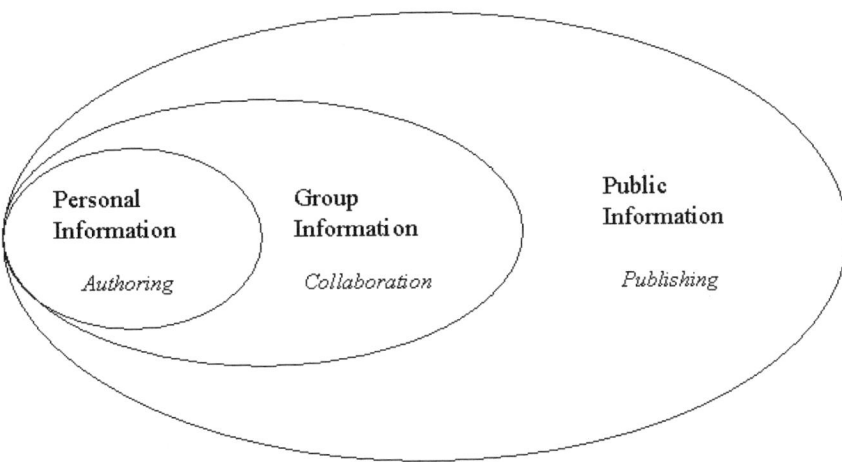

Figure 3.3 *Information concepts.*

viduals compose and edit their own presentations, spreadsheets, or other documents. With the advent of the Web and increased use of e-mail it became possible to publish these files and thus make them available on a broad scale.

More recently many applications have followed the lead of Lotus Notes and provided a means of collaboration between multiple users. While these are frequently departmental applications, there is no requirement of belonging to the same organizational structure. In fact, ad hoc groups have often sprung up to work on specific projects. Applications that are able to grant discretionary access in a simple and efficient way are better placed for use in such a dynamic environment.

3.3.2 Entertainment

Electronic books

Electronic books, or e-books, are books in digital format that can be read on all types of computers, in particular handheld devices. In addition to the text and graphics of printed books, they can embed audio and video or reference external content via hyperlinks.

Similar to traditional books they can be ordered through the mail or purchased in bookstores. But they can also be downloaded or received by e-mail. The electronic format reduces costs of shipping and storage and also removes constraints that have been imposed for broad audience appeal.

Video

Video is clearly a very popular form of entertainment. The challenge has always been reproducing it in a palatable manner on a mobile device. There are two aspects to this problem. One is the display, which must offer good resolution and luminosity. This can be difficult on a small device, which automatically limits the size of the display.

The other problem is that the content of good (high-resolution) video is large. Either the video must be loaded on the device beforehand, which means there must be sufficient local storage, or else the video must be streamed, which means high bandwidth.

Music

Mobile music devices have been around for a long time. Even before the Sony Walkman people had portable tape recorders. But the size has been continually reduced. Now the MP3 format has opened up the way to storing music in microscopic size. But even more importantly we are now able to play music on general-purpose systems. We do not need a dedicated MP3 player but instead can use a software implementation on any mobile device.

Instant messaging

Chatting was one of the first Internet applications to take off. Even before the Windows and the Internet I remember using VAX Phone to accomplish the same purpose using character-cell terminal. Since then there have been quite a few different standards and products offering users to communicate online in real time.

In many ways it could be argued that mobile devices, which are increasingly equipped with phone capabilities, do not have any need for typing messages. However, the success of Short Message Services clearly demonstrates that this is not the case. Many users are willing to type, even though the option of speaking is clearly available.

Games

No discussion of entertainment would be complete without mentioning games. They are difficult to justify with business reasons, but that does not stop many, particularly the younger population, from playing them almost addictively.

Wireless connectivity expands on the possibilities, since it facilitates multi-player games in real time without requiring the players to be physically together.

Application development

There is a common concept that redesigning data for mobile devices involves reducing these data to a smaller scale. This is certainly one aspect, but it oversimplifies the task. A laptop user will be only marginally more satisfied with four-line Web pages than a phone user trying to read a complex Web page intended for desktops.

In the future, content will need to be developed in a device-independent fashion and rendered to device-specific constraints using a scalable procedure. The first part of this is addressed by moving all data to an XML format. However, the second part is unlikely to be solved by simply creating a few style sheets. Careful planning will be required to create an architecture of XSL templates that is able to render the data of all applications to all client devices.

The advantage of developing applications that reside on the operating system is that they can also run in "disconnected" mode when the device has no connection, either wired or wireless, to the network. However, this must be weighed against the disadvantage of the lack of portability. Any applications developed for one operating system will not run on another without considerable rework.

3.4 Synchronization

The use of a local cache of server information is a great advantage in that it allows the user to benefit from the data when the machine is disconnected. However, it also implies an additional requirement—namely, the ability to synchronize the changes when a connection is reestablished.

Most synchronization solutions are specific to a platform (e.g., Palm or Windows CE) and application (e.g., Outlook or Lotus Notes). Some of the more prevalent solutions include the following:

- Intellisync (Puma)
- Hotsync (Palm)
- ActiveSync (Microsoft)
- Synchrologic
- Visto
- LINQ
- Avantgo

3.4.1 SyncML

In order to simplify this growing set of solutions and reduce the need to install overlapping and redundant software, an initiative was formed by some of the major vendors in this space (IBM, Matsushita, Motorola, Lotus, Palm, Nokia, Psion, Ericsson, and Starfish).

SyncML is a markup language based on XML that can convey synchronization requests in an abstract and vendor-independent manner. In order to achieve this it is also independent of both the underlying transport and the data that it encapsulates.

Independence of the transport protocol clearly implies that both SyncML devices need to agree on using the same one. But there are several available, including the following:

3.4 Synchronization

- HTTP
- WSP (i.e., WAP)
- OBEX (e.g., for Bluetooth or IrDA)
- SMTP, POP, or IMAP (i.e., via mail)

There is no restriction on the type of data that can be synchronized. However, in its definition it requires the inclusion of the following:

- Personal data (including the predefined formats of MIME, vCard, vCalendar, and iCalendar)
- Relational data
- XML and HTML
- Binary data

What SyncML does provide is a syntax for encapsulating synchronization data. SyncML messages specify whether a particular data item has been deleted, added, or modified. They also provide sync anchors (e.g., time stamps), which permit another device to assess whether the item has previously been synchronized.

Architecturally it is important to note that in any particular dialog one device is always considered the SyncML client and the other the SyncML server. It will often be the case that the client is a phone or PDA while the server is a PC. However, this is not required.

Synchronization types

A number of sync types have been defined that characterize a particular synchronization request. One or more of these may be requested in any given session.

Two-way sync—A normal sync type in which the client and the server exchange information about modified data in these devices. The client sends the modifications first.

Slow sync—A form of two-way sync in which all items are compared with each other on a field-by-field basis. In practice, this means that the client sends all its data from a database to the server and the server does the sync analysis (field by field) for these data and the data in the server.

One-way sync from client only—A sync type in which the client sends its modifications to the server, but the server does not send its modifications back to the client.

Refresh sync from client only—A sync type in which the client sends all its data from a database to the server (i.e., exports). The server is expected to replace all data in the target database with the data sent by the client.

One-way sync from server only—A sync type in which the client gets all modifications from the server, but the client does not send its modifications to the server.

Refresh sync from server only—A sync type in which the server sends all its data from a database to the client. The client is expected to replace all data in the target database with the data sent by the server.

Server alerted sync—A sync type in which the server alerts the client to perform sync—that is, the server informs the client to start a specific type of sync with the server.

Synchronization commands

Each synchronization type will translate into a sequence of SyncML packages, broken down into messages that consist of individual commands. SyncML defines the following "request" commands:

Add. Allows the originator to ask that a data element or data elements supplied by the originator be added to synchronization data accessible to the recipient.

Alert. Allows the originator to notify the recipient. The notification can be used as an application-to-application message or a message intended for display through the recipient's user interface.

Atomic. Allows the originator to indicate that a set of commands should be performed with all or nothing semantics.

Copy. Allows the originator to ask that a data element or data elements accessible to the recipient be copied.

Delete. Allows the originator to ask that a data element or data elements accessible to the recipient be deleted. A Delete command can include a request for the archiving of the data. The deletion can either be a soft or hard delete.

Exec. Allows the originator to ask that a named or supplied executable is invoked by the recipient.

Get. Allows the originator to ask for a data element or data elements from the recipient. A get can include the resetting of any metainformation the recipient maintains about the data element or collection.

Map. Allows the originator to ask the recipient to update the identifier mapping between two data collections.

Put. Allows the originator to put a data element or data elements on to the recipient.

Replace. Allows the originator to ask that a data element or data elements accessible to the recipient be replaced. This command makes a complete replacement of the data element. This command must only be specified within a Sync command.

Search. Allows the originator to ask that the supplied query be executed against a data element (or elements) accessible to the recipient.

Sequence. Allows the originator to indicate that a set of commands should be performed in the specified sequence.

Sync. Allows the originator to specify that the included commands should be treated as part of the synchronization of two data collections.

SyncML specifications

In addition to two protocol specifications (SyncML representation protocol and SyncML synchronization protocol) SyncML provides additional documentation and support to ensure interoperability and help implementors get started with the language. Some of the supplementary resources include the following:

- An architectural specification
- Bindings to common transport protocols
- Interfaces for a common programming language
- An openly available prototype implementation of the protocol

3.5 Extensibility

No computer can accommodate the entire set of needs of its user base out of the box. It wouldn't be efficient to make features and interfaces that are used only occasionally part of the standard configuration. The case is even clearer with mobile devices, where space and power constraints make it necessary to carefully plan the design of the appliances.

In order to augment the functionality of the equipment it is necessary to attach accessories. These may be part of a mobile kit, in which case they need to be miniaturized, or they may be part of a stationary connection point.

The most common connectivity to date has been via copper wires, such as serial or USB cables. Often the device can be attached to a cradle that has built-in connectivity. Increasingly, wireless connections are also being offered. The first of these to be used was Infrared, but there are also options available now to use Bluetooth, 802.11b, and cellular phone connections (GSM, IS-95 CDMA) and/or packet data networks (CDPD, GPRS).

A wider range of connectivity options becomes available if the device can support additional modules such as PC cards, CF cards, or Sony's memory sticks. There have also been innovative approaches of Handspring with its Visor and Compaq with the iPAQ Pocket PC jackets, which enable additional, but proprietary, forms of extensibility.

No matter what hardware is attached to these devices it is important to ensure that the corresponding software, usually in the form of drivers, is available on the mobile device.

3.6 Smart cards

The ultimate mobile device is neither a phone nor a PDA. Smart cards are available in form factors several orders of magnitude smaller. The development of these cards, spearheaded in France and Germany, has advanced greatly over the past decade. They now sport full operating systems, including CPU, storage, and I/O channels.

Their focus has traditionally been that of providing security. They provide an ideal storage for private keys, passwords, and cryptographic algorithms, since they are all but tamper proof. As their storage has increased, they have also begun to include sensitive personal information such as account numbers.

But we have not reached the end of the road regarding their development, since their power increases can be used to store nonsensitive personal information. Not only will they serve as personal identity, but they can store all manner of personal information.

With the inclusion of Bluetooth, smart cards can transparently interact with other (mobile and stationary) devices. If necessary they can function as the master store for personal data and synchronize changes with any peripheral appliances.

3.7 Voice processing

The ultimate goal is for the smart card to function as a wearable personal identity module. Whether included in a ring, watch, or necklace it can be maximally mobile and make use of other peripherals, such as keyboards, monitors, microphones, and speakers, to interact with the user.

Some of the interfaces to smart cards include the following:

- STK—The SIM ToolKit (STK) standard gives user-friendly access to mobile value-added services through the use of menu-driven services that are added to the handset menus. STK menus and services are stored on the SIM card.
- Windows for Smartcards
- Javacards

3.7 Voice processing

Conceptually voice processing is unrelated to wireless and mobility since it can be used across any medium and in any environment. However, it is wireless technology that is driving its use. The small form factor of mobile devices makes it difficult to provide them with a human-machine interface that can accommodate medium to large volumes of information. Since most of these devices are equipped with microphones and speakers, it is natural to consider them for voice enablement.

While in some cases ease of use is merely a benefit of voice-enabled devices, there are also situations where it is virtually a requirement. These include the following:

- Some countries are considering legislation barring the use of keyed input and graphic displays in vehicle-mounted systems due to the inherent danger of distraction while driving.
- Visually impaired people can compensate for their disability by speaking and listening with the same aptitude as those with full sight.
- Workers—for example, in construction—who require their hands to operate machines or need to focus their vision on particular points, are still able to communicate and interact with a computer while doing so.

According to IDC projections there will be 70 million smart phones by 2004. By the end of next year 98 percent of all smart phones will have some voice capability included. Phones currently have too small a processor to be able to provide sophisticated speech functionality. A phone or Palm PDA

Chapter 3

might be able to work with a vocabulary of 20 words and is mainly speaker dependent.

Some PDAs have more processing power than the Palm and are therefore able to offer more advanced solutions. An example in point is the Fonix solution, which demonstrates ASR and TTS on an iPAQ PocketPC. With its StrongARM processor it is a very competitive PDA in the voice market.

3.7.1 Distributed speech processing

In order to be successful, voice processing needs to be distributed. The client devices, although improving, do not provide sufficient power to completely process all voice information. Servers often have this capability, but it is not effective to have them perform all the processing due to the increased error rates and costs in passing voice data across a network.

With the need for distribution, and the desire to make all components interoperable, comes the requirement to develop a standard. VoiceXML (previously VoxML and now often called VXML) is developed under the auspices of the W3C and is available as a standard from their Web site. VoiceTIMES, led by IBM and Philips, is another initiative that is driving some complementary standards, such as audio capture.

3.7.2 Multi-modal user interaction

While voice input and output are ideal under some circumstances, it is equally clear that they are not suited for all situations. In airplanes, business meetings, and libraries it would be antisocial to begin a loud dialog with a computer. But even in the privacy of an office or home, voice is not always the most effective mechanism to convey information. Photographs and schematic diagrams are often difficult to put into words.

The term multi-modal refers to the ability of an application to respond to different modes of input and also provide multiple modes of output, as required by the user. In Windows we can compare this to the possibility to either use the keyboard (Alt-F) or the mouse (Click on the File menu) to obtain the same result. Adding voice means that the user can either speak a command or use an alternate input mechanism (such as a phone keypad).

The output could similarly be sent to the speaker but also to a display screen. In some cases it might not be the same information (it could be a diagram), but it should obviously be correlated. If the user turns off the speaker, the display still receives the output.

One often-cited type of multi-modal interaction is "Voice-In, WAP-Out." This would help WAP to overcome one of its biggest deficiencies, which is limited input on small form factor devices. This is only one example. The main idea is to make the options and combinations as open and flexible to the user as possible.

3.7.3 Automatic speech recognition

Probably the two most influential factors in determining the quality of speech recognition are the power of the machine doing the interpretation and the quality of the corpus (body of empirical language data the processor has available). A corpus must contain sample information reflecting the entire population (including dialects, accents, and different voices). It is interesting to note that the corpus grows directly as a system is used. In other words, most ASR systems have the capability to "learn."

Speech recognition can also be improved by using effective noise cancellation and echo cancellation techniques. Particularly in a driving vehicle these are very important factors, which can lead to significantly different levels of accuracy.

The most accurate systems today can reach levels of 98 percent in favorable circumstances and approximately 82 percent in noisy environments. This is very similar to (possibly higher than) the level of accuracy another person would expect to obtain. One major drawback is that people are more likely to realize when they didn't understand a word and can request confirmation. In ASR this is called "barging in," and some systems do support it; however, their basis for deciding when a word is ambiguous is mostly acoustic/phonetic, at best syntactic, and clearly not semantic.

3.7.4 Text to speech

Text to Speech (TTS) has improved significantly over the past few years from the metallic computer voice usually associated with it. There are currently two primary techniques used to synthesize speech, as follows:

1. Formant speech synthesis requires less hardware (although typically still 50 MIPS) and is lower quality. It is based on individual phones (as in phonetic elements—not telephones) or phone-pairs (diphones).

2. Concatenative speech synthesis sounds more natural, since it is based on word phrases. It requires more resources to implement,

making it difficult to install on mobile devices. But it is becoming the norm for server-based systems.

Additionally there is a technique called "phrase-splicing." It is the most natural sounding but is only suitable for applications with a very limited number of phrases.

E-mail is considered the application to most benefit from advanced TTS. It is clearly also a very ambitious effort when you reflect on the wide vocabulary, frequent misspellings, and ambiguous abbreviations that are common in many e-mail messages.

3.7.5 Speaker verification

Speaker verification through voice is a very effective authentication mechanism for mobile applications since it does not require a portable scanner or reader of any kind. It is also unobtrusive and relatively inexpensive to implement. There are several verification types that can ensure varying levels of accuracy by requiring users to present supplementary authentication data with the words they say. So, for example, by having the user supply a password it is possible to verify the password at the same time as the voice.

It might seem like it would be possible to trick a voice verification system by taping the voice and replaying it for authentication. However, this is not as easy as it sounds and can be detected by analyzing reverberation and correlation. In case of any ambiguity a second random prompt would easily unmask an impostor.

That is not to say that there is no possibility for error. While it is unlikely that an impostor would be clearly identified as the person he or she is impersonating, there are many situations where the verification is not able to determine clearly whether or not the voice belongs to a given person.

In summary, it would be very difficult to stage an intrusion. The mathematics prove that the likelihood is very low. Nevertheless, in order to be safe a second authentication factor is also recommended.

3.8 Management

So much for the theory. What does it take to implement applications using mobile devices? The first big decision is the selection of the devices. There are essentially two approaches to this.

The company may issue a decree that only one, or a small set, of devices will be supported. In this case the support overhead is reduced but individual user needs may be neglected.

On the other hand, if users are allowed to make their own choice, either individually or departmentally, they can ensure that all requirements are addressed. However, given the diversity of devices, the challenge of the IT department grows with each new model.

3.8.1 Deployment

After the selection of the devices comes the selection of the applications. A major factor to consider here is how the applications will be installed on the machines, not only for their initial release but for any subsequent updates.

One approach often taken for fixed systems such as desktops is for the IT department to physically visit each computer and install any new software. This is not likely to work with mobile users, who take their tools with them when they travel.

Another approach has been to use a central software system that performs any upgrades when the user logs onto the network. Again decentralized users may not be able to afford the long connection time over slow and expensive connections.

3.8.2 Backup and restore

An effective software deployment mechanism will ensure that applications can be reconfigured on a damaged or lost device. However, it does not help with critical user data that might be lost. There are at least two different approaches that can be adopted to address this need. They are as follows:

1. Users can synchronize all their data with a server that is integrated in a reliable disaster recovery solution.
2. Specialized solutions can directly store user data on recoverable media.

The latter solution will typically require more effort on the part of the user, but it may permit the user to restore even in situations where no server synchronization is possible. On the other hand, if a complete synchronization solution is already in place and software can be efficiently deployed to mobile users, then the combination of these two may already represent a satisfactory backup solution.

3.8.3 Support

Remote support has been available in many organizations even before the advent of mobile devices. However, the need has increased as more users require and expect round-the-clock, and often round-the-world, support for their devices and applications.

3.8.4 Tracking and monitoring

Most companies have mechanisms to track their hardware and software assets. Accounting procedures often require these mechanisms, and a complete and accurate inventory can help in supporting the users. Support can also be improved by monitoring the devices as well as the network performance and availability. Both of these tasks are much more challenging in a mobile environment where the devices are geographically dispersed and wireless networks often use foreign infrastructure, which is not directly visible.

3.8.5 Security

We will address the subject of security in more detail in Chapter 6. However, while on the topic of mobile devices it is important to realize that there are some very specific security needs for mobile devices that differ from the requirements of a fixed desktop.

Some specific areas that need to be considered include the following:

- Virus vulnerability—mobile devices are often used outside the corporate firewall and are therefore more susceptible to viruses. At the same time they are also not as powerful and therefore less able to scan for known viruses locally.

- Exposed passwords—passwords that are entered in public and/or are cached on the system provide hackers with an easy means of access to the corporate network and applications by observing password entry or stealing the device.

- Weak air interfaces—while most modern wireless interfaces have built-in security, they all have their weaknesses, which can be exploited by someone with enough determination.

3.9 Trends in mobile devices

3.9.1 Convergence of phone and PDA

As mentioned previously, there is a trend for the phone and the PDA to converge in functionality. Phones are increasingly adding functionality as they grow in processing power. At the same time the PDAs require connectivity and both voice input and output. This makes them capable of telephony with very little incremental effort.

This does not mean that we are likely to have a single, all-encompassing device in the future. It merely means that the differentiation will no longer be in terms of these two categories. However, there will always be different user needs, and vendors will honor them with different configurations.

An important factor is the size of the device. It is a very personal choice whether one wants a smaller, less-functional tool or a larger appliance capable of running more applications. Implied in these decisions are the size of the screen and the processing power (e.g., in terms of CPU speed and size of RAM) of the system. So we can expect to see continued diversity along those lines.

The choice of human-machine interface is also very individual. Whether input should be performed using a keyboard, a stylus, or voice will have very different implications for the design of the device.

And clearly the applications that are run will be different. These may not all come on the device out of the box, but some applications will have specific hardware requirements and many will simply not be made available on all devices.

3.9.2 Multi-modality

One specific trend in the human-machine interface is multi-modality. Rather than relying on one particular input and output mechanism users will be able to select the one that is most fitting to the circumstances.

Multi-modal systems are already familiar to users of desktop systems, who switch between the keyboard and the mouse seamlessly. On a mobile device the stylus might be used to point or to write. Alternatively a keyboard might be connected, through infrared, cable, or Bluetooth.

But increasingly one of the most effective means of interacting with a mobile device will be voice. As the processing power of the devices increases and they are able to efficiently recognize spoken commands, voice becomes the natural interface where it can be applied.

3.9.3 Connectivity

Many mobile devices have some means of connectivity already. However, they often rely on peripheral equipment for support. As connectivity becomes an indispensable feature of mobile devices it becomes more efficient to embed the transceiver in the device.

In fact, with the increasing number of air interfaces that might be suitable in different mobile scenarios, it is reasonable to expect the device to incorporate several different radios, which might be used as the user moves between coverage areas. For example, the user might require a Bluetooth or 802.11b connection in the office, but then switch to a GSM/GPRS connection on the road and possibly even a satellite connection when traveling to remote areas.

3.10 Summary

The variety in mobile devices presents a big departure from the standard desktops and servers that formed the foundation of the Internet. There is limited inverse correlation between the mobility ("smallness") of the devices and the functionality, or computing power, that they incorporate.

But the main item of diversity is in the machine interfaces that they use. Beyond keyboards and screens of different sizes they also use new mechanisms, such as a stylus or keypad and increasingly voice activation.

The PDAs are largely grouped in one of three camps depending on the platform they support: Palm, Windows CE, or EPOC/Symbian. On the other hand, mobile phones do not yet commonly use an open platform and can only be grouped by the browsers (WAP/WML or HTTP/HTML) they support and the machine interfaces they include.

As the computing power of the devices increases they will be able to run most desktop applications. But, with their current limitations, the focus is on personal and collaborative information, as well as entertainment, particularly in the consumer space.

3.10 Summary

Synchronization is increasingly important as multiple mobile devices replicate personal information on a variety of data stores. There are already a number of proprietary solutions addressing the issue but they provided limited interoperability between the full set of devices and applications. SyncML is an industry initiative with broad market adoption that attempts to provide a common platform for all future synchronization.

Mobile devices are limited in the functionality and connectivity they can provide out-of-the-box. In order to augment these for special needs it may be necessary to extend the platform with additional peripherals. This is facilitated with some of the units that offer more complete and ergonomic connectivity options than others. Many will offer Infrared ports but increasingly Bluetooth may be more available. PCMCIA and CF cards also provide alternatives as do serial and USB connections.

Voice processing is a natural match for mobile devices as a machine interface since it requires little device real-estate—it is demanding in terms of computational resources, however. As the units begin to incorporate faster processors, this option will become more prevalent, particularly as it lends itself to many of the mobile environments where the focused attention of keyboard entry is not possible.

Management of mobile devices, including deployment of the systems, development of a comprehensive backup and restore policy, tracking and monitoring, and remote support all represent a new set of challenges for IT departments. Many of these are inevitable, but their difficulty can be reduced if the users are encouraged to select devices from a small set of platforms using standard configurations.

The era of a standard working environment, where everyone works on similarly configured desktop systems, is fading into history. In its place we see a plethora of different devices, running different software, with different functions and machine interfaces.

Only time will tell whether this reshuffling will eventually stabilize and it will once again be possible to neatly categorize all devices, or, instead, whether we have reached a truly "personal" era, in which the set of tools begins to match the diversity in users' preferences and needs.

Either way, companies must now face the growing reality of increasing diversity in system requirements coupled with the recurrent need for mobility.

Bibliography and related Web sites

SyncML

White Paper: http://www.syncml.org/download/whitepaper.pdf
Standards: http://www.syncml.org/downloads.html

Murray, J. *Inside Microsoft Windows CE.* Microsoft Press, 1998.

Rhodes, N., and J. McKeehan. *Palm Programming.* O'Reilly, 1999.

Tasker, M. *Professional Symbian Programming.* Wrox Press, 2000.

PalmOS: http://www.palm.com/devzone/

EPOC: http://www.symbian.com/technology/v6-papers/v6-papers.html
http://www.symbiandevnet.com

Windows CE and Windows NT Embedded:
http://www.microsoft.com/windows/embedded/

Sun JavaOS: http://www.sun.com/javaos/

Microware OS-9:
http://www.microware.com/Products/Rsrcs/TechBriefs/OS9.html

Microsoft Mobile Explorer: http://www.microsoft.com/WIRELESS/

Smart cards

Rankl, W., and W. Effing. *Handbuch der Chipkarten.* Hanser Verlag, 1999.

PC/SC: http://www.smartcardsys.com/

PKCS#11: http://www.rsa.com/rsalabs/pubs/PKCS/html/pkcs-11.html

Opencard: http://www.opencard.org/

Javacard: http://www.javasoft.com/products/javacard/index.html

Gemplus:
http://www.tdap.com/tdap/wireless/wireless(gemplus_9903).html

Windows for Smartcards:
http://www.microsoft.com/windowsce/smartcard/

Javacards: http://java.sun.com/products/javacard/, www.javacardforum.org

4

Infrastructure

This chapter consists of two parts. The first part looks at the types of networks that are typical of wireless scenarios. Wireless configurations are usually part of a wireline infrastructure that encompasses most of the complexity needed to serve a large area with many users. However, this dependency is not absolutely necessary. This section includes an overview of the common topologies currently in use as well as a perspective on the kinds of networking configurations that are possible within a wireless environment.

The second part of the chapter investigates some of the frameworks and layered protocols that are available to facilitate wireless applications above simply the air transmission. In the future, open mobile platforms will probably provide reliable IP-connectivity to the applications and they will do the rest. However, this is not yet the case for many wireless networks and mobile platforms. This section will look at what is currently available.

Most wireless applications do not directly work with the air interface to interact with the clients. That would be an unnecessary burden on the programmer, who is mostly concerned with the business logic of the system than with how it is delivered to the end-user.

Using the rudimentary stack suggested previously, I have expanded my wireless model to include two additional layers, as shown in Figure 4.1.

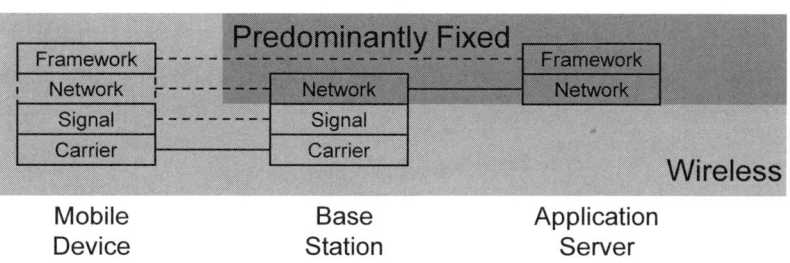

Figure 4.1
Addition of the network and framework layers.

I have added Network, which connects the base station to the application server, and Framework, which can be used to abstract the lower-level details from the application. The dashed lines refer to virtual, rather than actual, connections.

The application server could, of course, also connect to the base station with a wireless link. To keep the diagram simple I have ignored the lower layers of the application server, which could be implemented with either fixed or mobile connectivity.

Figure 4.2 illustrates what the term Network implies. Rather than one managed network between the transceiver and the application there are at least three networks to consider: the mobile operator's network (in the case of WWAN and public WLANs); the Internet at large; and the corporate network, where most of the infrastructure is likely to reside. They may use any number of network types from Ethernet LANs to optical fiber, or even other wireless links. The point is that collectively they provide connectivity between the base station and the application server.

Unless the client devices and/or application server are expressly dedicated to one application, then most of the common software is also part of the available frameworks. The only parts that are application specific are the application software and the data on both the client devices and the application servers.

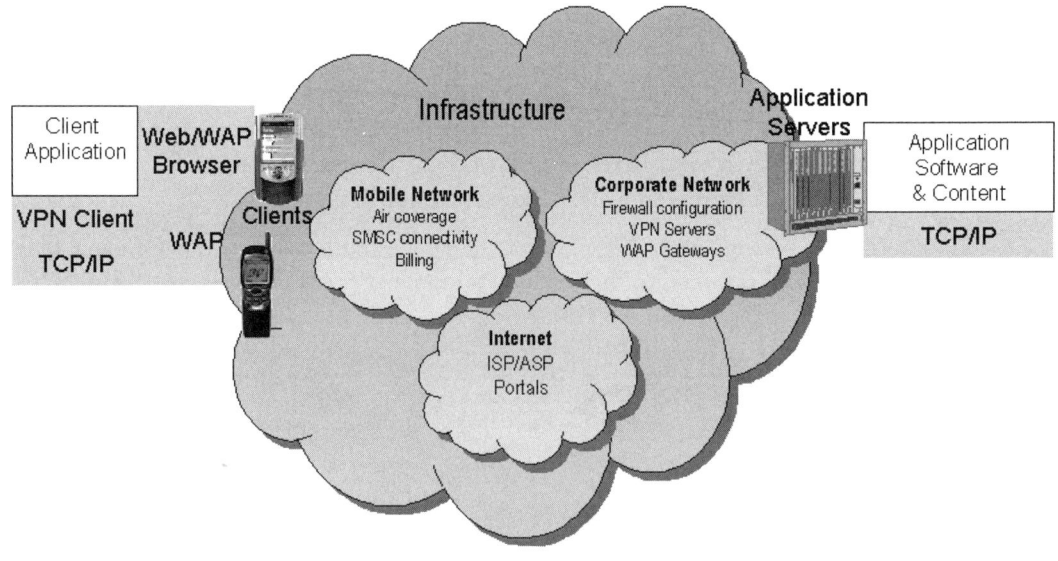

Figure 4.2 *Network layer explained.*

The mobile device is not necessarily aware of any network. It may simply make a connection to the base station. And the station would then represent the device to the network. There is, however, a need for the client portions of the framework to be installed on the device, since these are the components that actually interact with the user and make the application available.

At the other end, the application server must have some kind of network connectivity and must also accept the framework that is being used by the client. It might be more accurate to call this server a framework server, since it would be conceivable to separate the framework from the application itself. But that is an implementation detail and is not really within the scope of this discussion. For the time being, we will assume that we are dealing with one server that incorporates both.

4.1 Networks

As just mentioned, the network connectivity from the device to the application servers may involve a number of different networks and network types.

Figure 4.3 gives an example of a combination of wireless and fixed networks that might work together to provide a comprehensive solution.

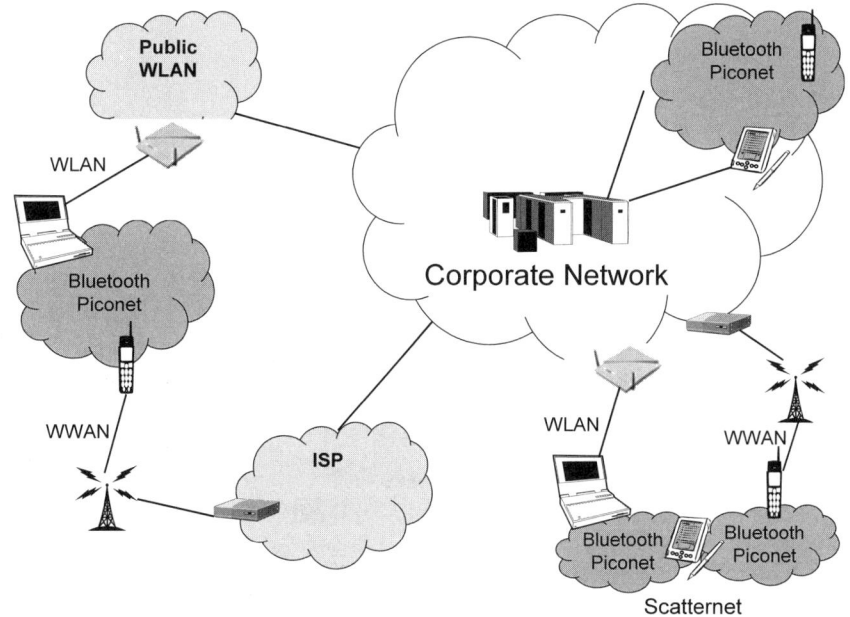

Figure 4.3   *Combining wireless and fixed networks.*

Clearly many other permutations would be possible. This is just to give you a taste for the potential complexity.

What is actually implemented will vary greatly depending on the needs of the users. At this point we are not yet ready to go into detail about network planning. We want to focus on the types of networks that are available at the wireless end. Let us go back to the figures from Chapter 1 that show the primary types of networks. (See Figure 4.4.)

Fundamentally, you have the option of connecting devices as peers. In this case each device encompasses the whole stack described previously. Each device can simultaneously function as a client, a receiving station, and an application server.

The usefulness of this kind of network is obviously reduced to the content each contributing device can provide. But the advantage is that it can be set up quickly and easily.

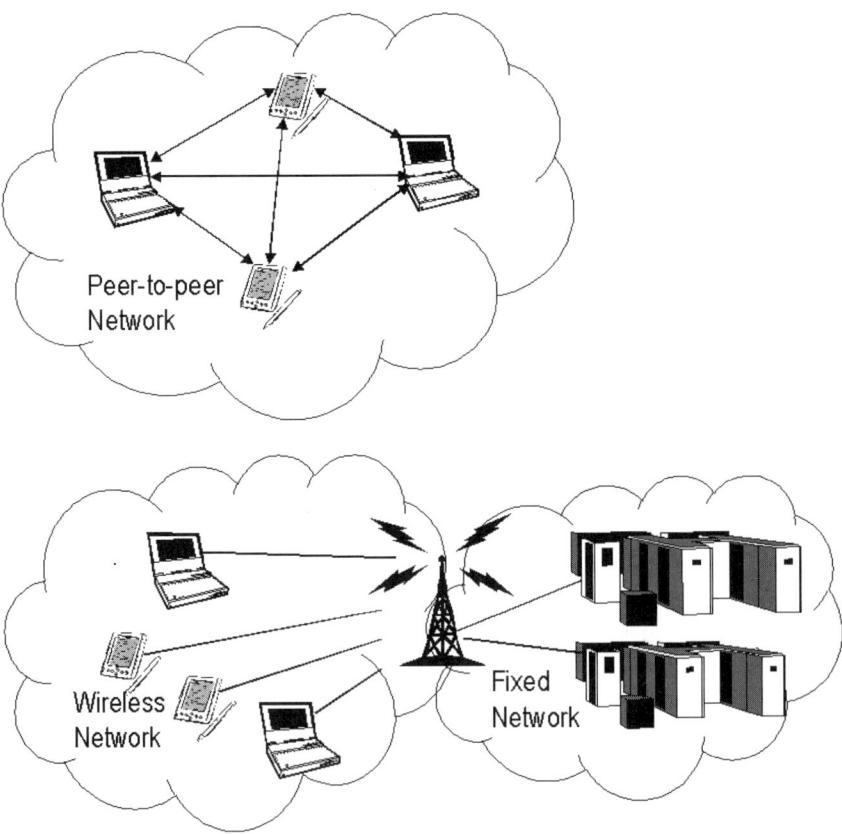

Figure 4.4
Peer-to-peer, wireless, and fixed networks.

The typical wireless network will be similar to the example at the bottom of Figure 4.4. There will be a fixed access point, which will relay all wireless access for a given area to the fixed network. Clearly this entails some overhead and additional planning in terms of investing in and optimally placing the base stations. But it provides devices with virtually the same access as you can obtain on fixed networks.

Ad hoc networks tend to be very simple. But as you can see in Figure 4.5, they do not necessarily need to be small. Bluetooth allows a maximum of eight devices (one master and seven slaves) to join the basic topology, called a piconet. However, it is possible to extend the total set of connected devices by merging multiple piconets. This requires that some devices participate in more than one piconet. Each one of these bridges must be the master in one of the piconets.

It is unlikely that this would grow to comprehensively cover any large area. The specification was not designed with this purpose in mind, and it is not tuned for large networks. However, this is a convenient mechanism for situations where more than eight devices must be connected in an ad hoc fashion.

Figure 4.5
Ad hoc network.

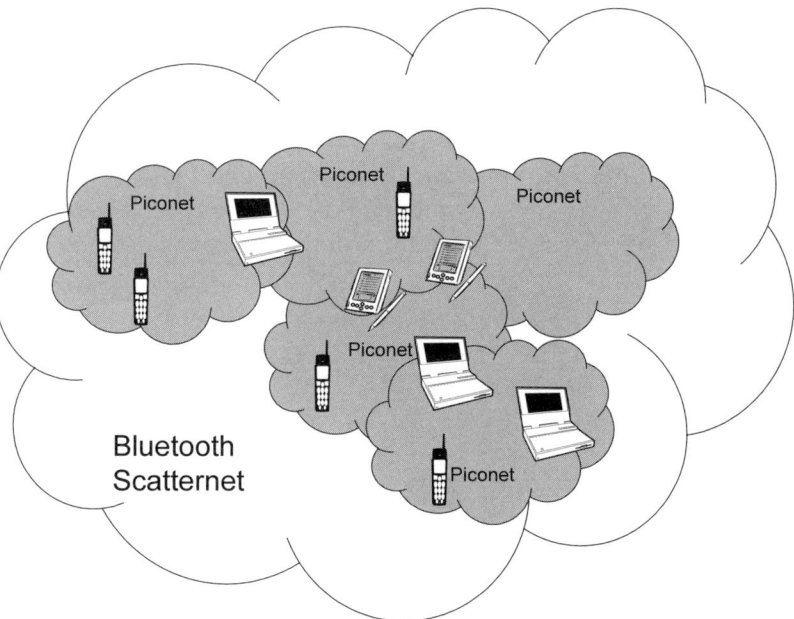

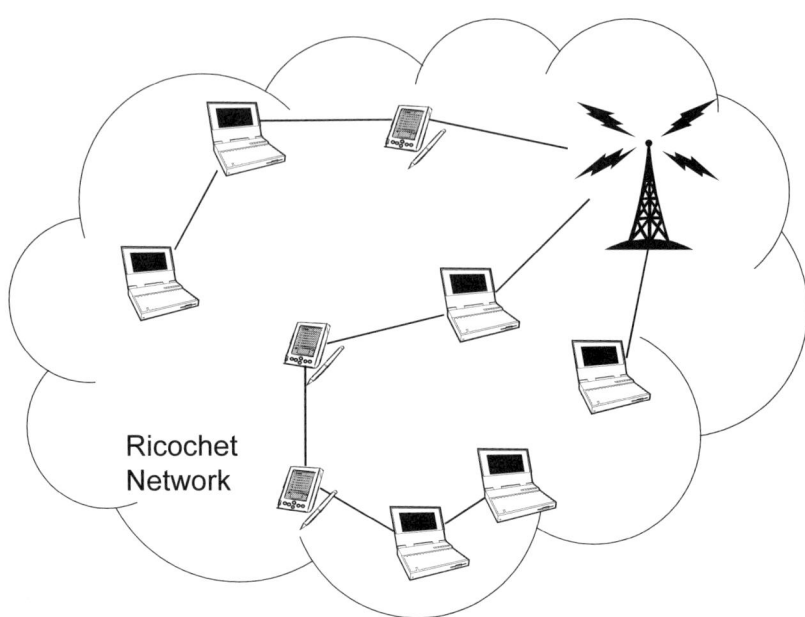

Figure 4.6
Ricochet network.

A Ricochet network (see Figure 4.6) is a novel approach at extending the concept of the base station reach. Rather than requiring all the devices to connect directly to a base station, they can instead connect to other devices, which are (directly or indirectly) connected to the fixed transceiver.

This approach can be used to drastically improve the range of the base station. For example, in theory you could have four evenly spaced devices extending in a line from the access point. The last device could then be four times as far away from the base station as the range would normally allow. (See Figure 4.7.)

Figure 4.7
Four evenly spaced devices in a line from the access point.

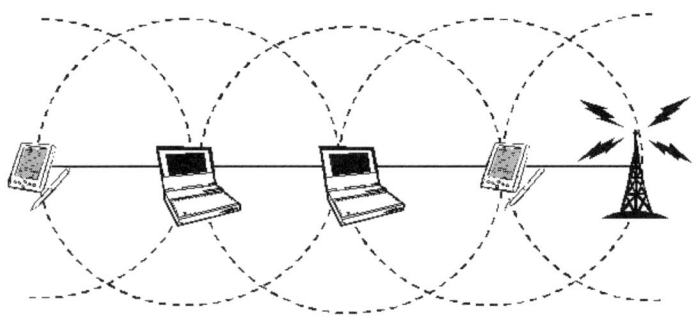

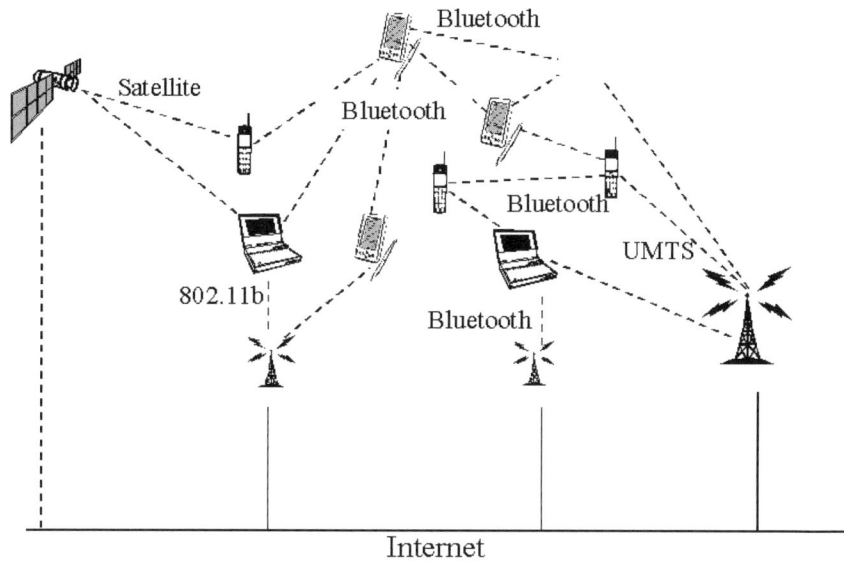

Figure 4.8
Multiple air interfaces provide optimized coverage.

4.1.1 Heterogeneous networks

As more air interfaces become available, we can expect increasing diversification in the types of networks and topologies that enable wireless access.

Figure 4.8 gives an illustration of how multiple air interfaces can complement each other to provide optimized coverage of a particular area. The miniaturization of the radio components makes it possible for each device to have multiple standards. And the increasing use of IP for connectivity makes it possible to transparently switch between networks to improve performance and reduce costs.

4.2 Frameworks

The next few sections will discuss some of the best-known frameworks that exist and that provide a platform on which to develop and deploy wireless applications. These frameworks include the following:

- SMS
- WAP
- i-mode

- IrDA
- Bluetooth

This list is not exhaustive, but it does cover much of the areas receiving most attention today. SMS is an application in and of itself. However, it also provides support to other layered applications. This section will deal with its supportive role. WAP and i-mode both offer similar types of functionality, but have taken a different approach to the problem of delivering Web content to small mobile devices. IrDA is not a radio-frequency protocol, however, it is wireless and provides much of the framework on which Bluetooth is based.

There is no norm as to what constitutes a framework. The category is very informal and includes technologies at different levels and with different constraints. For example, WAP was developed with the goal of being able to work over virtually any air interface. The Bluetooth architecture, on the other hand, implies a Bluetooth air interface.

We should also take care not to view these frameworks as operating entirely in the same space or providing the same functionality. They are not necessarily competitive. In fact, in some cases they can work together. WAP can use Bluetooth as a bearer for its services, for example.

One last point to keep in mind is that there is no absolute need for any of these as a prerequisite to wireless applications. After a short overview of each of these technologies, we will look at how networking is evolving with packet-switched data and see what impact this might have on the further development of these frameworks.

4.3 SMS—Short Message Service

Short messages are a primitive form of data transfer supported by several cellular phone standards (IS-136, IS-95, and iDEN) but used primarily by GSM. The latter allows text messages with a length of up to 160 single-byte characters to be sent to other GSM subscribers. Virtually all GSM operators and handsets support SMS.

Over 100 billion SMS messages are sent between mobile phones every year, and the number is growing rapidly. Yet, while SMS is actually a very successful application in and of itself, it is not easy to use for long messages or custom applications. Its main attraction is that it is universally available to GSM subscribers and can, therefore, be employed by other products and technologies as a delivery mechanism for other information. (See Figure 4.9.)

4.3 SMS—Short Message Service

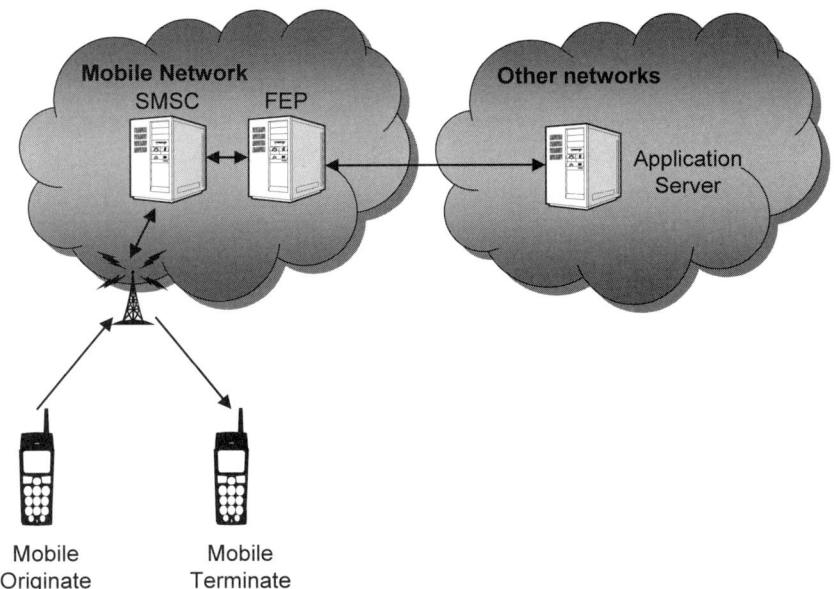

Figure 4.9
SMS can be used as a delivery mechanism for other information.

4.3.1 SMS applications

Several handsets supporting Personal Information Managers (PIMs), such as the Nokia Communicator, support direct interaction between SMS and the PIM. An application can send an e-mail message, business card, or calendar appointment embedded in an SMS message and the user can import these into the phone's PIM with a single click.

The challenge is for the application to send and receive SMS messages. There are two fundamental approaches to this, as follows:

1. Use of a GSM/SMS modem on the server or LAN
2. Direct connection to the SMS Center

SMS modem

As an example of the first option, Xsonic uses a GSM modem connected to the network to transfer data via SMS to mobile devices. These data can be e-mail notifications (Xsonic InTouch) or corporate data (Xsonic DataNow).

Other configurations would include GSM PCMCIA cards or even internal or external modems insofar as they are capable of GSM/SMS.

Direct connection to SMS Center

Many mobile operators also allow direct connections to the SMS Center (SMSC). There is no industry standard interface but there are several SMSC vendors, and most supply some interface. Some of the more common vendors include the following:

- CMG Telecommunications
- Nokia
- Ericcson
- Motorola
- Lucent
- Comverse Network Systems
- Logical Aldiscon
- Sema Group

Direct connections have the advantage of being faster and more reliable than mobile-originating SMS messages. However, they require both the agreement of the mobile operator for direct connections and also that the application supports the protocol in use by the operator.

Fenestrae's Mobile Data Server makes direct connections to SMSC, using several supported protocols, in order to expedite the delivery of SMS messages.

4.3.2 Future of SMS

As newer wireless interfaces, such as WAP, appear on the market, the long-term future of SMS is questionable. In the short term, however, it is unique in the two following areas:

1. A wide user base—while only a fraction of phones in the GSM market are currently WAP-enabled, almost all are able to take advantage of SMS. This means that SMS-based applications have a much larger potential market than WAP applications over the next year.

2. Notification—SMS is inherently push oriented, while browsers, such as WAP, are pull oriented. While it is simple to simulate pull functions with push commands, the converse is not true. In the medium term WAP 1.2 browsers using packet data networks such as GPRS will also be capable of notifications. However, these are

currently not available and will not be universally adopted for some time.

There are also efforts underway to extend the functionality of SMS so that it may become useful for a new range of applications.

Some of the potential enhancements to SMS may include the following:

- Long Message Services (LMS)—It may be reasonable to remove the 160-byte limitation from SMS and allow longer packets to be exchanged. However, it is already possible to transmit larger amounts of data by segmenting then into SMS packets. A much more useful improvement would be a reduction in SMS latency and developments on the devices to be able to automatically process a multi-packet data stream.
- Multimedia Messaging Services—In addition to text there is an increasing need to transmit other types of data, such as streaming voice and video. Again, it is quite possible to use current SMS as a simple packet data network to transmit opaque information. However, the true benefit will only be reaped when there are standard formats that all devices and applications can understand.

4.3.3 The lesson from SMS

Many corporate users discount the use of SMS given the difficulty they have in entering text on a mobile phone. This complaint is common with all mobile devices. It is certainly easier to type in a message on a keyboard than a phone keypad. Intuitively, it seems as if the technology could not be viable for input on a wide scale.

Reading a message is not quite as cumbersome, even though it is usually also limited by a small display. But it is quite acceptable for short messages. And, in fact, there are more Mobile Terminate (MT) SMS messages, composed either by an application or a PC user, than there are Mobile Originate (MO) SMS messages. The surprise is that the number of MO SMS messages is still extremely large (in some countries they average more than one per user per day) and is growing at a fast pace.

A look at the user base reveals that these are not typically business customers. It is the consumer market, in particular teenagers and even preteens, who have wholeheartedly embraced the new technology. Possibly they have more time to enter text or less money to obtain access to traditional Internet devices.

But much more fundamentally, they are willing and able to quickly adapt to new interfaces. Some of them can enter data on a mobile phone at mind-boggling speeds. They are not dissuaded by the challenge of something new and can see its potential in a less biased light.

4.4 WAP

4.4.1 Architecture

The intent of WAP is to provide a framework within which applications on the wireless device can function similar to a desktop browser.

One criticism leveled at WAP is that it has developed its own protocol stack rather than making use of and influencing the Web protocols already in use. The main reasons why it was considered necessary were as follows:

- To provide an acceptable user interface for smaller devices, such as phones
- To enable communication over a high-cost, low-bandwith, and unreliable network connection

The easiest way to understand WAP is to look at the similarities and differences between HTTP/HTML and WAP/WML.

Figure 4.10 shows the functional approximation of some of the protocols.

WAE

Wireless Application Environment (WAE) is a collective term that includes both WML and WMLScript.

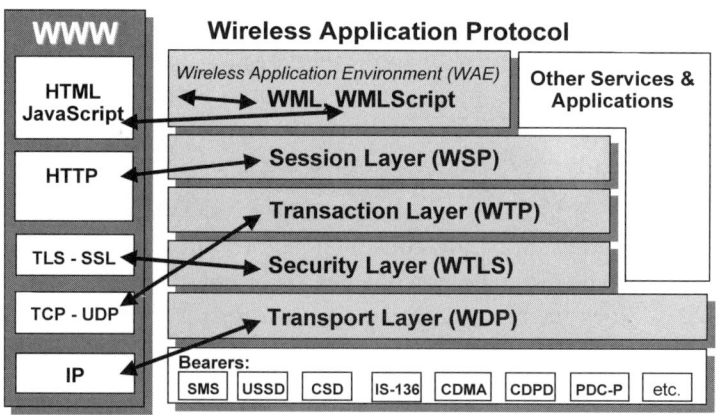

Figure 4.10
WAP protocols.

WML

Wireless Markup Language (WML) is shown in the following code segment.

```
<?xml version="1.0"?>
<!DOCTYPE wml PUBLIC "-//WAPFORUM//DTD WML 1.1//EN"
"http://www.wapforum.org/DTD/wml_1.1.xml">
  <wml>
    <card>
      <p>
       Hello mobile world!
      </p>
    </card>
  </wml>
```

The origin of WML lies in HDML—Handheld Device Markup Language—developed and deployed by Openwave (previously called Phone.com and Unwired Planet), which offered it to the WAP forum as a starting point for a wireless markup language. Since HDML is still a dominant phone browser in North America, many Openwave gateways will translate WML 1.1 for execution on HDML browsers.

WML has much in common with HTML. They contain many of the same tags and both adhere to a similar syntax. One important, albeit subtle, difference is that WML is based on XML, whereas HTML is based on the more lenient SGML. This implies a stricter syntax as follows:

- Only lowercase is permitted.

- References to Document Type Definitions (DTDs) are required.

- All elements must be terminated: For example, A <p> element must be followed by </p>.

In version 2.0, the plan is for WML to merge with XHTML, which is also based on XML.

A second difference in the markup languages is that HTML pages are both a unit of packaging and also a unit of navigation. WML uses the terms deck and card to distinguish between these two concepts, as shown in the following code segment.

```
<wml>
  <card>
    <do type="accept">
      <go href="#card2"/>
    </do>
  <p>Press OK to display the next screen.
  </p>
```

```
        </card>
        <card id="card2">
        <p>This screen displays the content of Card 2.
        </p>
        </card>
    </wml>
```

Deck

A deck is a unit of packaging typically stored in one file. A WSP request will fetch one deck in order to optimize the number of requests. One deck can include more than one card.

Card

A card is a unit of navigation. It may be larger than the display area, but it is still considered the unit of navigation. In other words, the user interacts with one card at a time.

Since cards in the same deck are downloaded simultaneously, it is possible to transfer quickly between cards. No request from the server is necessary.

In the previous code segment, the first card displays a message and sets up an action to occur when the "accept" key is pressed. That action is to go to the second card.

WMLScript

The wireless application environment also offers a scripting language similar to JavaScript, as shown in the following code segment.

```
function priceCheck(givenPrice) {
   if (givenPrice > 100) {
      var newPrice = givenPrice;
   } else {
      newPrice = 100;
   };
   return newPrice;
};
```

WMLScript adds some capabilities that are not supported by WML, such as the following:

- Checking the validity of user input
- Accessing facilities of the device and allowing the programmer to make phone calls, send messages, and add phone numbers to the address book

- Generating messages and dialogs locally, thus reducing the need for expensive round-trips to show alerts, error messages, and confirmations
- Allowing extensions to the device software and configuring a device after it has been deployed

WML Script is loosely based on ECMAScript (from which JavaScript is derived). Some of the Advanced features were dropped, and it was optimized for low bandwidth and minimal memory requirements.

WSP

Wireless Session Protocol is referred to as WSP.

HTTP is a stateless protocol. This means that a Web server does not need to cache any information about a user accessing its site. Instead each HTTP request supplies any context information necessary to reestablish a session.

This is an advantage for the Web server, since it reduces the amount of resources it needs in order to scale to thousands of concurrent users. However, it also implies that a substantial amount of information needs to be supplied with each request.

Over wireless links the need to reduce data transfer is considered more important than the need to limit server load. WSP is, therefore, not a stateless protocol. Instead, the gateway caches header information to avoid the need for retransmissions. This also makes it possible to suspend and resume long-lived sessions.

In order to further reduce the transmission size, content types and headers are encoded in binary form.

WTP

Wireless Transport Protocol (WTP) is derived from TCP but optimized for narrowband, high-latency networks. Some of its characteristics include the following:

- Segmentation and reassembly
- Concatenation of protocol data units to reduce number of packets
- Delay of acknowledgments in order to concatenate them with other data

WTLS

Wireless Transport Layer Security (WTLS) is based on TLS (TLS is based on SSL V3.0 and expected to eventually supplant SSL). It provides data integrity, privacy, authentication, and denial-of-service protection.

In order to establish a secure session a full handshake is required to establish a secret symmetric bulk encryption key. This key is then retained across connections in order to reduce the necessary interaction.

To conserve bandwidth WTLS also supports compact (ANSI X9F1) public-key certificates in addition to X.509 certificates.

WDP

Wireless Datagram Protocol (WDP) is roughly equivalent to IP in that it offers datagram support. This is the least common denominator of all environments. WDP ensures a common interface to datagrams regardless of the underlying technology.

If the underlying bearer does not provide segmentation and reassembly, the WDP provider interface will implement it in a bearer-dependent way.

In the long term, IP will probably be used as a routing protocol in most networks (e.g., in GPRS). In those cases WDP can and will use UDP. Even in the short term the most common bearer is Circuit-Switched Data (CSD), which means that a circuit is established with the WAP gateway (e.g., using a PPP connection) and that this circuit will typically support IP and thus UDP.

There are, however, other means of delivering the data to the WAP gateway. It is possible to use SMS; however, the delays involved in sending an SMS message for each request and waiting for an SMS message in response are usually prohibitive.

Other bearers have also been defined in the WAP specification; however, they have not yet seen widespread use. Note that the set of bearers is not closed by the specification. Other technology, such as Bluetooth, may also provide interfaces from WDP.

WTAI

The Wireless Telephony Application Interface (WTAI) framework supports wireless telephony applications that interface with the in-device telephony-related functions and the network telephony infrastructure. It extends the WAE by adding an interface from WML and WMLScript to a specific set of local, telephony-related, functions in the client. This interface is called the Wireless Telephony Application Interface.

WTA services are created using WML and WMLScript. From a WMLScript, telephony functions can be accessed through the Wireless Telephony Applications Interface (WTAI). WTAI also provides access to telephony functions from WML by using URIs. While the WTA services reside on the WTA server, the client addresses these services by URL.

Examples of WTA services include the following:

- Incoming call selection—The service is started when an incoming call is detected in the client. A menu with user options is presented to the user. Examples of options could be the following:
  - Accept the call
  - Redirect to voice mail
  - Redirect to another subscriber
  - Send a special message to caller

- Voice mail—The user is notified that he or she has new voice mails and retrieves a list of them from the server. The list is presented on the client's display. When a certain voice mail has been selected, the server sets up a call to the client and the user listens to the selected voice mail.

- Call subscriber from message list or log—When a list of voice, fax, or e-mails or any kind of call log is displayed, the user has the option of calling the originator of a selected entry in the list or log.

4.4.2 Topology

While the WAP stack is modeled on the Internet/Web stack, it is fundamentally incompatible with it at most levels. Since WAP relies on the Internet to store most of its (WML) content, this is a dilemma.

In order to bridge the two worlds it is necessary to implement a gateway (called WAP proxy or WAP gateway), which performs the protocol translation from WAP to HTTP/TCP/IP. (See Figure 4.11.)

A WAP session begins when a customer selects the appropriate option on a WAP-enabled handset. The request is encoded using the WAP protocol stack and sent via the Network Operator to a predefined WAP gateway. There the request is interpreted and passed on to a particular URL via HTTP/TCP/IP.

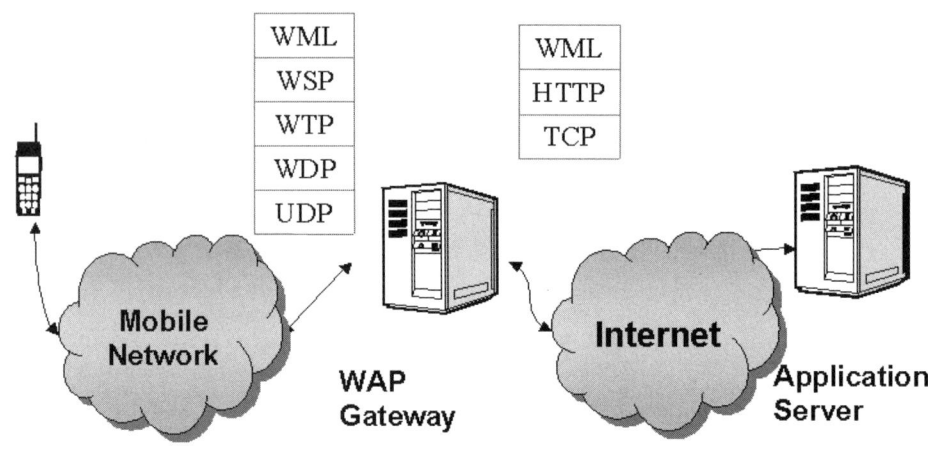

Figure 4.11   *WAP gateway.*

This model is sufficient when the mobile terminal has IP connectivity—for example, using a packet data network such as GPRS.

Most current implementations of WAP use circuit-switched data, such as GSM. They require a dial-up server (such as Microsoft RAS services), which will provide IP over point-to-point protocol (PPP). (See Figure 4.12.)

4.4.3   Is WAP dead?

The assertion that WAP is dead is not uncommon, but it means different things to different people. In order to address the assertion we need to look at the various criticisms.

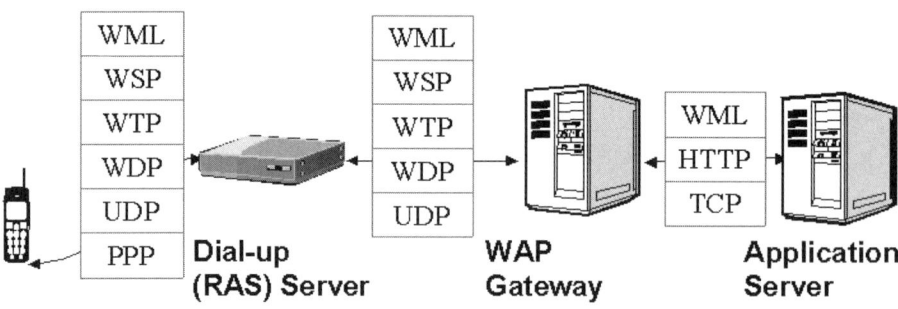

Figure 4.12   *WAP requires a dial-up server.*

Overhype

There are those who say it is dead because of the disappointment the customers experienced when trying to use it. There is no question that many mobile operators and phone manufacturers overhyped the standard. They implied that it was possible to attain a typical Internet experience through a WAP-enabled phone, which is clearly not realistic. As a result the public is now more than a little skeptical. But their next phone will still probably be a WAP-enabled phone, since they will soon have no other option—whether they use WAP is, of course, another story.

Divergence from Internet standards

A more technical criticism is that WAP should never have been designed and that instead the phones should be using HTTP over TCP/IP. There is some merit to this criticism; however, there are also some very good reasons why this approach was not taken (the handset manufacturers and operators wanted to have a standard that would work over poor networks and on small form factor devices with no common transport layer). Regardless of whether it was the best decision the reality is that the number of WAP devices will be growing exponentially over the coming years, since all major phone manufacturers endorse WAP (and the big players endorse *only* WAP).

Human-machine interface of small form factor devices

A substantial amount of skepticism exists as to whether there are any meaningful applications to push to the phones. Part of the overhyping, mentioned previously, also included the misconception that people would be reading documents, browsing complex Web sites, and even entering verbose textual data, on the phones or other small form factor (SFF) devices. This is not likely to happen. Phone-based applications will involve highly targeted information, such as alerts, which don't require much or any user entry. These applications are only now starting to appear. Many of the more interesting ones for phones include "push," which is currently not really feasible with WAP. However, GPRS and WAP 1.2, which are now becoming available, will facilitate this.

Dual-modal devices (e.g., using Voice-in and WAP-out) might also facilitate the input of data for SFF devices, while WAP could still be the optimal protocol for output.

Poor performance

The long delays of setting up circuit-switched connections and the limited bandwidth available in 2G networks have made the process of browsing WAP sites even slower than would be expected given the awkward data entry mentioned previously. As we move to 2.5G and 3G networks with higher bandwidth that are "always on," these problems should be reduced. In fact, browsing WML pages should be much faster than browsing HTML pages, since they contain much less information.

Competing technologies

In addition to WAP there are other possible technologies that can be used on mobile devices: Regular HTML browsers or i-mode. Many press articles have implied that i-mode might have better success, since it has shown phenomenal adoption rates in Japan. However, we must be very cautious in extrapolating this to other parts of the world. Any other approach will have many of the same challenges as WAP and will have the disadvantage that it will be a late start. On the other hand it may be able to bypass some of the "WAPlash" by employing some more realistic marketing.

The current status

WAP has had a slow start, both in the devices and the applications. It is starting to gather some momentum technically and must now be considered on its technical merit.

The current projections are that WAP will gradually merge into the HTTP stack. With more reliable and better networks the only valid reason for keeping it would be for WML, which caters to small form factor devices. However, WML 2.0 is also expected to merge into XHTML.

Most of the effort in redesigning an application for WML is to migrate to an XML model. Once that is done, the actual syntax of the markup language is almost a trivial factor.

4.5 i-mode

i-mode is a proprietary technology developed by NTT DoCoMo, one of Japan's leading mobile operators. It is currently only used in Japan, although NTT DoCoMo has shown some interest in attempting to export its success to other markets. It supports a full range of information services offering everything from news and restaurant guides, to movie reviews, horoscopes, and a railway network navigator.

4.5 i-mode

While there is nothing to stop it being used on other platforms, its target is the phone. It requires these devices to be configured with a special browser, which has been developed by NTT DoCoMo.

The idea behind i-mode is to use the HTML language and HTTP protocol as much as possible. However, there are two major factors to consider. Phone displays are smaller than typical monitors, and, therefore, they are not suitable for receiving the bulk of the HTML content available today. Furthermore, they do not have the same processing power and are used in situations with less than ideal network connectivity. These issues should sound familiar to you, since they were an integral part of the WAP discussion. They are recurring themes and the base of much of the development in the wireless realm.

The Japanese approach to the problem was to develop a compact version of HTML called cHTML. While it is technically different from the industry standard WAP, the look and feel are very similar. It has some cosmetic difference, particularly in the area of graphics, where it offers color, that have made it very appealing to the consumer market.

i-mode resides on top of a packet data network (see Figure 4.13). In the case of Japan this is DoCoMo's proprietary PDC-P, which runs on the Personal Digital Communications (PDC) 1,500 MHz air interface. However, if and when i-mode is exported to other countries it will need to run on other packet data networks such as GPRS. Even in Japan it will soon be available on UMTS.

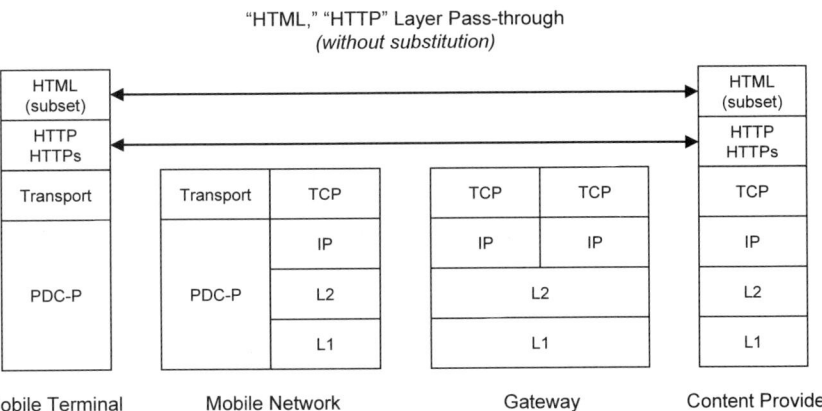

Figure 4.13 *i-mode's network system.*

i-mode makes use of the common Internet protocols and formats. In particular it directly uses the following:

- HTTP over TCP/IP as the transport protocol to request and deliver any content
- HTML, or rather a subset thereof, as the markup language of the content
- SSL as the end-to-end security provider between the mobile device and the application

As such it is partially compatible with existing content. The most obvious difference is that cHTML is only a subset of HTML, and it is therefore necessary to avoid the use of some of the tags. cHTML also introduces two new tags that are particularly relevant to mobile phones:

- Any of the special phone "access keys" can be specified as a trigger to a link. When the user presses the key, the destination is activated. For example: <A href=http://www.nttdocomo.com/ accesskey='1'> NTT DoCoMo</A>
- A phone number can be specified as the destination of an href. A click on this link will then dial a voice call. For example: <A href="tel:09044448888">Call now</A>

It is also important to recognize that i-mode itself is in evolution. Some of the earliest terminals do not support SSL and Java, for example. This means that an application might not want to provide the same content to all i-mode devices. Instead it will also need to examine the HTTP_User_Agent header before delivering data.

4.5.1 Is the i-mode boom exportable?

At present 365 companies are providing information services through NTT DoCoMo's own portal, which appears as default every time users start the i-mode function on their phones. An additional 7,000 sites are available from other companies and individuals. i-mode has been increasingly attracting subscribers (currently over 24 million, according to the NTT DoCoMo Web site on May 2001, barely two years after being launched), as the number of services available via the handsets grows.

This provides a stark contrast to the slow uptake of WAP in roughly the same time period. NTT DoCoMo has been the subject of many rumors contending that they will export the i-mode solution to other countries where they can expect similar success.

It is hard to exclude any possibility in the uncertain world of wireless. However, there are several points that might explain its exceptional success in Japan and some challenges that might mitigate its success elsewhere. These points are as follows:

- Part of the appeal of i-mode has been the fact that it is always on and can receive pushed content. These are possible since i-mode uses a packet data network. In other areas packet data networks are also being introduced, which will enable i-mode use but will also provide the same benefits to other technologies, such as WAP.

- i-mode is a proprietary solution. This means that it can be modified by one single entity. The flexibility is an advantage in an emerging technology, since it can quickly adapt to user needs and does not require consensus in order to be defined. However, this flexibility can become a liability if it is not used wisely. It can also serve to exclude other parties from integrating their requirements. For many this is not an issue. After all i-mode is almost the same as the standard Internet protocols, but it is not exactly the same.

- Many of the challenges and criticisms of WAP outside Japan have been centered around the small form factor of the device. i-mode does not change the form factor and may therefore receive the same criticism in other areas of the world.

The ultimate question is whether i-mode is truly superior to WAP, either in its technology or marketing. Or whether instead it benefited from favorable circumstances (Japanese culture, existence of a packet data network) that explain its success.

4.6 IrDA

IrDA was the first attempt at wireless cable replacement. Its objective was to allow simple ad hoc connections to be established without having to physically connect wiring. This could be useful for data transfer between mobile devices, such as notebooks. It could also be used to connect mobile with fixed devices—for example, to print out a document from a PDA.

IrDA supports high-speed, short range, point-to-point data transfer. The standard only specifies that connections must be supported up to one meter, but most products will connect at greater distances. Since all connections are one to one and line-of-sight, the physical security implied is greater than that of diffuse infrared.

IrDA is specified as a protocol stack with four required layers, as follows:

1. Physical Layer—Defines the optical characteristics, encoding of data, and framing for various speeds.
2. IrLAP (Link Access Protocol)—Establishes the basic reliable connection.
3. IrLMP (Link Management Protocol)—Multiplexes services and applications on the LAP connection.
4. IAS (Information Access Service)—Provides an index of services on a device.

In addition to the required protocols there are also several IrDA-defined optional protocols. Their use depends on the particular application. They are as follows:

- TinyTP (Tiny Transport Protocol)—Adds per-channel flow control.
- IrOBEX (OBject EXchange protocol)—Facilitates transfer of files and other data objects.
- IrCOMM—Emulates serial and parallel communications ports.
- IrLAN (Local Area Network access)—Provides infrared LAN access for laptops and other devices.

While infrared (IrDA in particular) is in widespread use today, it is losing ground to radio frequency devices, which are considered more reliable, are able to cover larger ranges, are able to pass through obstacles (such as walls), and support better mobility.

Nonetheless some of the protocols may continue to be used by the developing RF technologies (such as IrOBEX by Bluetooth).

4.7 Bluetooth

Bluetooth takes the concept of cable replacement a step further. It reduces the need for a line-of-sight connection. This means that the devices can be in movement without risk of losing the connection. It also means that they can connect at greater distance, and across physical obstacles, such as walls.

The types of cables that could be replaced with Bluetooth include the following:

- Serial (RS-232)
- Parallel

- Universal Serial Bus (USB)
- Mouse
- Keyboard
- Monitor
- Network

In general, any short wire can probably be replaced with Bluetooth, as long as it does not have very high bandwidth requirements (SCSI, IEEE 1394).

Bluetooth is designed to be integrated into very small devices without increasing the power requirements substantially. One of its energy-conserving features is that it allows the devices to enter four different levels of power consumptions, called Active, Sniff, Hold, and Park. The scheme is very elaborate and optimizes the amount of transmission that is needed when a device is temporarily inactive.

As we mentioned earlier in the chapter, the fundamental topological unit of Bluetooth is a piconet (set of up to eight devices in synchronized communication), which may be combined with other piconets to form a scatternets. Within each piconet one device must act as a master and control the air interface. Additionally there may be up to seven active and 255 parked slave devices. A scatternet, therefore, consists of multiple overlapping piconets (one device can be a slave of multiple piconets or slave of one and master of another).

One of the challenges for Bluetooth is to ensure that the binding process between devices is both simple and secure. It is not an easy task to make the binding of a mobile phone to the owner's headset transparent while at the same time guaranteeing that it will not bind to the headset of the owner's neighbor. This difficulty is compounded by the fact that some of the devices (e.g., headsets) do not lend themselves easily to advanced configuration, since they have no inherent means of directly manipulating settings.

4.7.1 Protocol stack

Bluetooth is defined as a stack of protocols, as shown in Figure 4.14. Not all of these protocols are Bluetooth specific, although they are all included in the architecture.

The core protocols are as follows:

- Baseband

- LMP
- L2CAP
- SDP

Audio—The audio model is relatively simple in Bluetooth; any two Bluetooth devices can send and receive audio data between each other just by opening an audio link.

LMP—The Link Manager Protocol is responsible for link setup between Bluetooth devices. This includes security aspects, such as authentication and encryption (by generating, exchanging, and checking of link and encryption keys), and the control and negotiation of baseband packet sizes. Furthermore it controls the power modes and duty cycles of the Bluetooth radio device and the connection states of a Bluetooth unit in a piconet.

L2CAP—The Bluetooth logical link control and adaptation protocol (L2CAP) adapts upper-layer protocols over the baseband. It provides connection-oriented and connectionless data services to the upper-layer protocols with protocol multiplexing capability, as well as segmentation and reassembly operations.

SDP—The Service Discovery Protocol facilitates queries of device information, services, and the characteristics of the services. Based on the results, a connection between two or more Bluetooth devices can be established.

RFCOMM—The Cable Replacement Protocol emulates RS-232 control and data signals over Bluetooth baseband, providing transport capabilities for upper-level services (e.g., OBEX) that use a serial line as a transport mechanism.

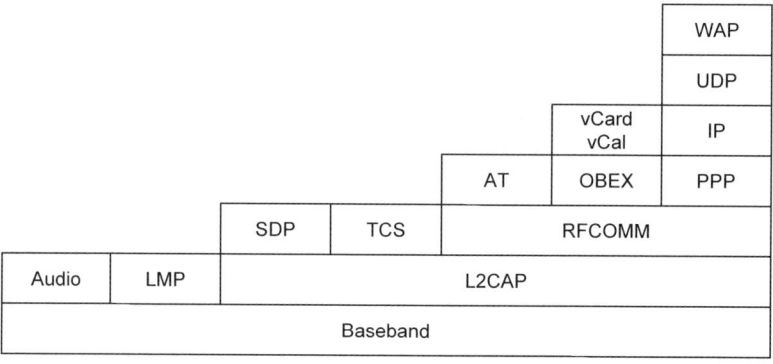

Figure 4.14
Bluetooth protocol stack.

TCS BIN—Telephony Control Protocol Binary defines the call control signaling for the establishment of speech and data calls between Bluetooth devices. In addition, it defines mobility management procedures for handling groups of Bluetooth TCS devices.

AT Commands—The Bluetooth SIG has defined the set of AT commands by which a mobile phone and modem can be controlled in the multiple use models.

4.7.2 Profiles

In addition to the protocols discussed previously, Bluetooth also defines a set of profiles. These facilitate a standard means for applications on various Bluetooth devices so they can interoperate. As illustrated in Figure 4.15, the profiles are partially dependent on each other. For example, the Dial-up Networking Profile requires the Serial Port Profile, which, in turn, builds on the Generic Access Profile.

Generic access profile

The purpose of the Generic Access Profile is to perform the following:

- Introduce definitions and common requirements used by the transport and application profiles
- Describe the standby and connecting states in order to guarantee that links and channels always can be established between Bluetooth devices
- Describe discovery, link establishment, and security procedures
- State requirements on user interface aspects

Service discovery application profile

The number of services that can be provided over Bluetooth links may increase in an undetermined (and possibly uncontrolled) manner. The Service Discovery Protocol (SDP) assists the user in identifying the variety of services available. It locates services that are available on or via devices in the vicinity of a Bluetooth-enabled device. These services are then returned to the user, who may elect to use one or more of them.

Cordless telephony profile

The Cordless Telephony profile defines the protocols and procedures that are used by devices implementing the "3-in-1 phone," which uses Bluetooth

Figure 4.15
Bluetooth profiles.

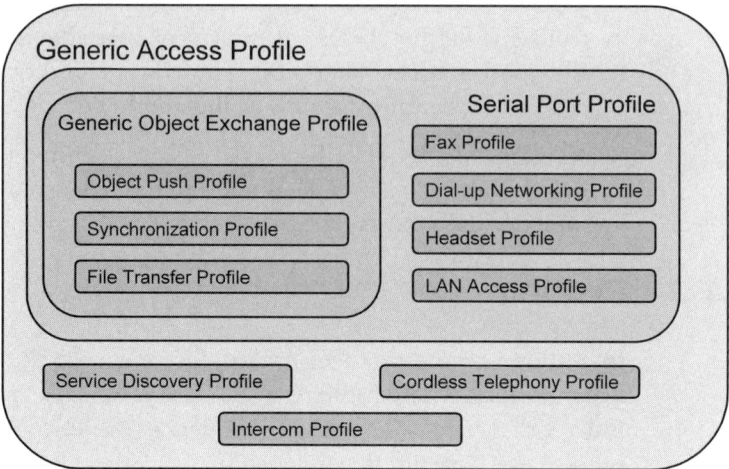

as a short-range bearer for accessing fixed network telephony services via a base station.

Intercom profile

This Intercom profile defines the protocols and procedures that are used by devices implementing the intercom part of the use model called "3-in-1 phone." More popularly, this is often referred to as the "walkie-talkie" use of Bluetooth.

Serial port profile

The Serial Port profile defines the protocols and procedures that are used by devices using Bluetooth for RS232 (or similar) serial cable emulation. The scenario covered by this profile deals with legacy applications using Bluetooth as a cable replacement.

Headset profile

The Headset profile defines the protocols and procedures that are used by devices implementing the "Ultimate Headset." The headset can be wirelessly connected for the purposes of acting as a PC or cellular phone's audio input and output mechanism, providing full-duplex audio. The headset increases the user's mobility while maintaining call privacy.

Dial-up networking profile

The Dial-up Networking profile defines the protocols and procedures that are used by devices implementing the "Internet Bridge." This includes the use of cellular phones and modems by computers, both to receive data calls as well as to establish dial-up connections to Internet access (e.g., RAS) servers.

Fax profile

The Fax profile defines the protocols and procedures used by cellular phones and modems to provide wireless fax functions to requesting computers.

LAN access profile

This profile defines the provision of LAN access to one or more devices over RFCOMM using PPP. It also describes how to enable PC-to-PC connections using PPP networking over serial cable emulation.

Generic object exchange profile

The Generic Object Exchange profile (GOEP) defines the protocols and procedures used by applications that need object exchange capabilities. Examples include PCs, PDAs, and mobile phones that require synchronization or file transfer.

Object push profile

The Object Push profile makes use of the Generic Object Exchange Profile for object exchange. It covers the process of pushing an object (such as a business card or appointment) to another device.

File transfer profile

The File Transfer profile makes use of the Generic Object Exchange Profile for object exchange. It defines procedures for browsing and manipulating the object store of another device. This includes navigation of the folder hierarchy, transfer of objects (both files and folders), and the direct manipulation of objects—for example, to delete files or create new folders.

Synchronization profile

The Synchronization profile defines the requirements for the protocols and procedures used by the applications providing synchronization. It defines the procedure for exchanging PIM—Personal Information Management data, such as phone book and calendar items—between Bluetooth devices. The data transfer must include the necessary log information to ensure that the data contained with the respective object stores are made identical. This synchronization can be initiated either automatically (when two devices are in RF proximity of each other) or manually at the user's request.

4.8 Pure IP networks

Current wireless technologies are based on a variety of different network types. Wireless local area networks (such as 802.11b) are primarily IP-based, but most wide area networks rely on circuit-switched data or proprietary packet data systems. As we evolve to a world where all data communications are realized with IP, the model of wireless solutions is subject to change, and the added value of some of the frameworks mentioned earlier in this chapter will diminish.

In particular the dividing line between common infrastructure and specific applications becomes much more evident. At present many solutions are tightly bundled with the air interface and supporting network. This obfuscates the functionality of the individual components, and thus makes it very difficult to pick and choose modules or to fairly compare the total value of two competing solutions.

This division falls along the following lines:

- Infrastructure must provide a reliable and secure IP connection from the mobile device to the network where the application is resident. This might be the corporate network, in the case of an enterprise. But the same concept also applies to consumer services. For example, it might be a secured banking network for mobile financial services.

- Applications should be independent and agnostic of the networking technology they use for their connectivity. This does not mean that all legacy applications are suitable for mobile use. However, it does mean that the same software must be optimized for high-speed wired and wireless access just as much as it is for dial-up on a phone line or for use on an unreliable and slow wireless network.

There are several different wireless technologies that can provide IP connectivity from the mobile unit to the application-hosting network, and many more are in development. Bluetooth, 802.11b, CDPD, GPRS, and the emerging 3G standards will support packet data. Eventually maybe even satellite communications will be IP based.

There are two common services that infrastructure can add to the networking technology, as follows:

1. Compression—the air interface is a valuable resource, certainly in time, but also often in cost (air-time charges). It should be optimized as much as possible. Some applications may compress the channel between the mobile device and the server application, but

many will not and instead will rely on the infrastructure for this support.

2. Security—some of the air interfaces are public, or hosted by a foreign entity. Even when the entire network is privately owned, the medium is inherently vulnerable and must be protected.

The best approaches, depending on the needs, of dealing with both of these services is through the use of a VPN or end-to-end encryption such as SSL/TLS. The client must be installed on the mobile device and the server must be attached to the edge of the target network. The current challenge to this approach is that most mobile devices do not yet have a VPN client available for them. However, this is changing and will probably not be a long-term problem.

4.9 Summary

In this chapter, we looked at the two primary wireless topologies: Those which connect ad-hoc/point-to-point and those which connect via an access point to a fixed network. However, we also saw, using Bluetooth and Ricochet as examples, that it is possible to extend these constellations in many ways in order to maximize the coverage of a pure wireless network without additional fixed infrastructure. In the future, combinations of these two may be possible with the primary wireless interfaces to enlarge the range of each network and therefore also the connectivity-choices of each mobile user.

Wireless frameworks allow mobile devices to run applications even though they have not completely adopted the Internet model of HTTP/TCP/IP. The most prevalent and successful framework is SMS which serves both as a primitive packet data network and a simple messaging system. WAP has received more publicity than SMS but has not been able to deliver the experience it promised. Nonetheless it is widely endorsed and provides a means of accessing reduced Internet-like content from most of the mobile devices being sold today. i-mode provides similar functionality to WAP but has managed to sell itself much better. Although it is currently limited to the Japanese market it has been extremely successful there.

With the advent of truly open mobile platforms, including phones running EPOC or Windows CE, and the replacement of current wireless systems with IP-routable networks, the need for intermediate networks will be reduced. The job of the network will be to provide secure, reliable and efficient end-to-end connectivity between the user and the application. It will be up to the client and server applications to optimize the user experience using the common IP transport.

Bibliography and related Web sites

Ricochet

Metricom: http://www.ricochet.com/

SMS

GSM World SMS overview: http://www.gsmworld.com/technology/sms.html

i-mode

NTT DoComo http://www.nttdocomo.com/

Eurotechnology i-mode FAQ: http://www.eurotechnology.com/imode/faq.html

Mobile Media i-mode FAQ: http://www.mobilemediajapan.com/imodefaq/

WAP

Mann, S. *Programming Applications with the Wireless Application Protocol.* John Wiley & Sons, 2000.

Arehart, C. et al. *Professional WAP.* Wrox Press, 2000.

WAP Forum: http://www.wapforum.org/

Bluetooth

Official Bluetooth web site: http://www.bluetooth.com/

Bluetooth Special Interests Group: http://www.bluetooth.org/

Miller, A. et al. *Bluetooth Revealed.* Prentice Hall, 2000.

Bray, J., and C. Sturman. *Bluetooth: Connect Without Cables.* Prentice Hall, 2000.

5

Applications

The entire wireless infrastructure is useless unless we have applications that use it. The search for the killer application, which would justify investment in mobile technologies, has often been compared with the quest for the Holy Grail. It is elusive and it is questionable whether there ever will be a single application that can justify the investment.

Instead, it is the composite of both existing and potential applications, which act as the incentive. But before we actually begin to discuss the applications, we need to be sure that we understand what it means to enable them for wireless and mobile use. It might seem intuitive but there are many different ways that the task can be approached.

This chapter begins with an overview of the distinctions between mobile and traditional applications, as well as a broad categorization of approaches that can be used to make applications work over wireless networks. It covers some of the common mobile applications and focuses on e-mail to demonstrate the many different ways it can be enabled for mobility.

The chapter concludes with a look at how future applications can be optimized for wireless access so that they do not need an enabler to operate in a mobile environment.

5.1 Why are mobile applications different from traditional applications?

It would be convenient if all the applications that have been developed over the past years could simply be dropped into a mobile environment and continue to work without any additional effort. There are two main reasons why it is not quite that easy, both of which were briefly mentioned in Chapter 3. These reasons are as follows:

1. Many mobile platforms are closed.
2. There is much more diversity in the machine interfaces of mobile platforms.

A closed platform makes it impossible to install the application on the device itself. However, beyond that, it also often imposes a particular type of connectivity and may even require a specific intermediate technology. If the phone only uses WAP, then your only choice is to use WAP or not use it at all!

The mobile applications also need to have a way to present content and receive input from a wide range of devices. This can be very challenging, especially since the number and types of available client devices are continually changing, with new products coming on the market all the time.

Both of these reasons must be addressed in order to allow a wide range of mobile users to access the application.

5.2 Evolution of applications

Some of the most interesting wireless applications are clearly applications that have been developed specifically to take advantage of some of their mobile features. Over time, we will probably see many programs that fall into this category. But that is not where we should look to get started with wireless. The implementation of the technology is better represented as a phased approach, beginning with our current environment and ending in optimized, value-added wireless applications, as follows:

1. Enable legacy applications.
2. Develop applications catering to mobile users.
3. Add functionality specific to mobility.
4. Extend into the high-bandwidth streaming applications.

There are already many legacy applications that can be enabled for mobile access today. They may not necessarily be especially suited for wireless networks, but they have the great advantage that they are already deployed and have demonstrated value to their users.

Some of the applications that fall into this category include Web access, e-mail, and PIM tools. Some less-common tools include sales-force automa-

5.2 Evolution of applications

tion and access to corporate databases. This software is mature and does not need additional functionality to be valuable. But it can increase its value tremendously by being available wherever the user is.

As the current set of applications is extended to wireless users, many new types of applications will begin to appear that are written with wireless networks and mobile devices in mind. Later they will even begin to take advantage of wireless technology to add additional value. Some of the factors that can be leveraged include the location of the user, the identity of the user, and the fact that the user is almost permanently connected.

As technology advances, bandwidth increases, and networks become more reliable, we can also expect to see applications that take advantage of the improvements in the infrastructure to transfer large amounts of information—for example, for audio and video streaming and for real-time games.

5.2.1 Components of mobile applications

As illustrated in Figure 5.1, there are two main components of mobile applications: The presentation of the content to the user and the actual application data. The distinction between these two is the basis for the two main categories of mobile applications: Client-side presentation and server-side presentation.

The best solution for a given scenario will depend on a variety of factors, including the volatility of the data, the computing power and storage capacity of the mobile units, and the network performance.

Client-side presentation

In the case of a special-purpose device or an open platform it is possible for the client to take care of all the presentation. This has the advantage that

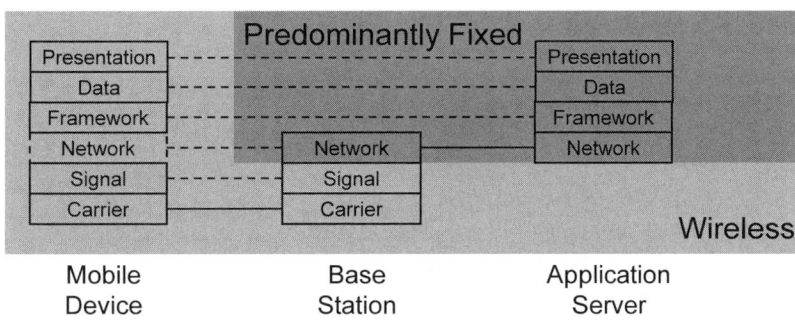

Figure 5.1 *The two main components of mobile applications.*

the content is always rendered in an optimal manner, since the client application "knows" its machine interface characteristics.

With this approach the server application merely passes important data to the client as the data are requested. The server does not need to concern itself with how the information is displayed to the user. The network load depends on how often and how much new information is requested by the user.

This approach is best suited for the following environments:

- The mobile units have significant storage capacity.
- Data do not change frequently.
- There are many different device types in operation.

Server-side presentation

The alternate approach is for the server to detect the end-user's device and render the content in a device-specific manner. Rather than only the application data it also sends display information to the client, which only needs to output it to the screen. In this case the server performs the bulk of the work. The client device does not have any application-specific components installed on it. Instead, it only has a general-purpose display application (such as a browser).

Server-side presentation is ideal for the following:

- Less powerful or unconfigured devices
- Rarely used applications
- Large user bases without sophisticated deployment mechanisms

Trade-off

Some of the trade-offs between the two approaches are displayed in Table 5.1.

Clearly client-side application implies a heavier load on the client in terms of storing the application. Server-side presentation is an additional load on the server, but since servers are not constrained in the same measure as mobile devices the incremental load is not an overriding consideration.

The deployment effort can be significant for client-side applications since every application must be installed on every device. This implies much more work than having one general-purpose display engine on each unit.

Table 5.1 *Trade-offs between Client-Side and Server-Side Presentations*

Criteria	Client-Side Presentation	Server-Side Presentation
Client load	–	+
Deployment effort	–	+
Development effort	+/–	+/–
Portability	–	+
Network load (static data)	+	–
Network load (volatile data)	–	–
Off-line usability		

The development effort of a specific version of each program for every device can be challenging. However, it is counterbalanced by the fact that the alternative is no less daunting, since it means not only detecting each device from the server, but also producing content that is tailored to it. A much more influential factor is the fact that applications rendered on the server can be supplied to a wider range of device types, including closed platforms, as long as they have some means of displaying general-purpose information.

The network load is highly variable for both of these approaches. With static data a client application may be able to reduce the load by redisplaying repeated information without resorting to the network. On the other hand, when data fluctuate frequently, this is less feasible and a server application has the benefit of only displaying the information that was actually requested by the user.

One final, but very important, factor is the usability of the application when there is no network available. This is only possible with a client-side application that can operate on local data.

5.3 Enabling applications for wireless access

Eventually there will be many applications that are particularly suited for wireless access. Designing them to take advantage of the strengths of typical mobile technologies will be an important factor. But to get started we need to look at how existing applications can be made to operate with a wireless infrastructure. Intuitively, we might think that there is one ideal way to tailor the applications so that they can be used in a mobile environment.

5.3.1 Access methods

In actuality, after some examination, we see that there are actually many different options, including those shown in the following chart.

Client-side	*Server-side*	*Both*
Native	Web	SMS
Enhanced	WAP	Proprietary
	Terminal services	

Native

Most applications will actually work over wireless links. Unless they are very sensitive to latency, require very high bandwidth, or implement poor error recovery, they should work over both circuit-switched connections and packet data networks. The applications are unaware of the network that lies underneath them. However, it is important to realize that the response time might be slow and the connections unreliable.

The option is the simplest to implement but it is also the least appealing to users with most current wireless networks. It does work satisfactorily over wireless LANs and will eventually be acceptable over 3G networks.

Enhanced

The most notable problems with native mode are the poor performance of the network and possible security issues when operating over public networks. It is possible to impose an additional transport layer that optimizes and/or secures the connection between the device and the application layer and is transparent to the application itself.

Web

Many applications have an HTTP/HTML interface available to them, which may not necessarily be the preferred native mode. In cases where these are distinct, the Web interface also presents an optional means of access. Similar to native mode it may also be enhanced for security and/or network performance.

WAP

This is clearly an option that is available and even geared to wireless access. It takes care of the security and network performance problems, albeit at the price of a rich user interface.

Terminal services

These services, made popular by Citrix, also represent a means of running applications on a mobile device. As long as the device has the client installed, it can run an optimized connection to the terminal services server that connects to the application server.

SMS

This is another wireless-specific interface. It is a powerful notification tool given its wide reach of SMS-accessible devices. But beyond end-user notification it is possible to use SMS to launch commands from a mobile device. Data can also be transmitted to phones/PDAs, which might automatically update a PIM based on SMS command conventions.

Proprietary solutions

These solutions, such as the RIM Blackberry, are also an important option. Since they can be tailored to very specific application needs and network infrastructures, they can be highly optimized and secured while still providing a high level of functionality and usability. This comes at the expense, however, of interoperability and extensibility to other applications.

To illustrate these we will go through each of these mechanisms and how they could be implemented with one particular solution (e-mail) later in this chapter.

5.3.2 Selection criteria

There are three main areas of selection criteria that help us to evaluate the available options for a given environment. We need to look at the set of networks, devices, and applications that will be used.

Network

The starting point for network selection is the area of coverage. The first choice is fundamental: What ranges do we need to be able to accommodate? Will it be possible to run purely on private wireless LANs—they are cheaper and offer higher bandwidth than any of the WAN options?

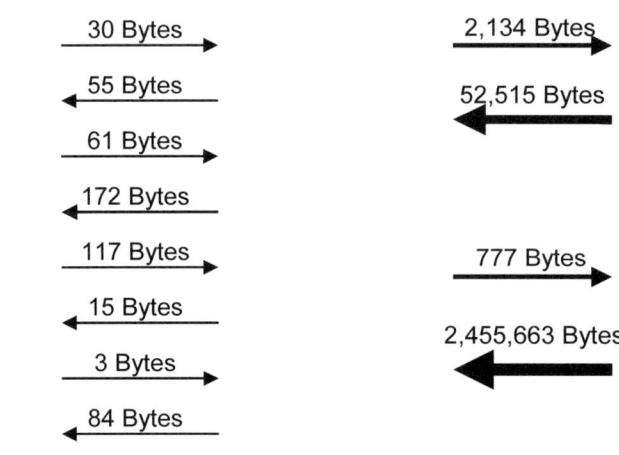

Figure 5.2 *Latency sensitivities.*

If we do decide that a public WAN or LAN is necessary in order to accommodate the complete area we wish to cover, then we can determine which service providers are able to provide our coverage the best. This decision may very well impact the choice of device, since not all devices can offer all networks.

Additionally, the performance of the network in terms of bandwidth, latency, and reliability is something that needs to be factored into any solution. As illustrated in Figure 5.2, some applications pass data in small chunks and are therefore very sensitive to latency. Since each request typically waits for the previous response, the whole interaction is exposed to numerous round-trip relays, which, cumulatively, can have significant impact on perceived performance.

Others have poor recovery features and are therefore incompatible with unreliable connections. They all consume some amount of bandwidth, so usability will be impacted by a slow connection. However, some of them make use of the existing bandwidth much better than others.

Device

As we already discussed in Chapter 3, there is a wide range of mobile devices with many different human-machine interfaces. These interfaces directly impact the choice of application access method, since they impose restrictions that are often incompatible or awkward with some of the input/output mechanisms.

Notebook

A notebook uses a very traditional and consistent machine interface. The screen is often a little smaller than on a desktop system, but it typically offers very high resolution, which partially compensates for this reduction. It also offers a full keyboard and an integrated pointing device.

The fact that it so closely resembles the traditional PC is a great advantage when it comes to using applications. Virtually all applications that will run on a typical desktop will also run on a notebook.

The main distinction between the notebook and the desktop is not in the power or functionality of the device but in the environment where it is used. Desktop systems typically enjoy good network connectivity and almost direct connections to the applications. Notebooks are used in a variety of different scenarios involving very slow links and circuitous paths into the corporate network.

Another important item to mention is that notebooks, similar to desktops, do not typically include any phone functionality. Their communications are limited to data transfer across the network, and they do not support voice or even some of the mobile phone data functionality, such as SMS.

In order to provide any type of wireless connectivity it is necessary to either include a wireless PCMCIA card or connect the laptop to a mobile phone—for example, using the serial port or Infrared.

Phones

Phones represent the other end of the spectrum. They have an extremely limited machine interface. While the interface is continually improving (with WAP phones, clamshells, and smart phones), it will probably always be constrained by the amount of space a small device can offer.

Most phones currently available offer very limited graphics. They are typically character-cell based, similar to the terminals that proliferated in the 1980s but with a much smaller keyboard. Similarly they do not offer a pointing device but instead rely on arrow keys or the equivalent.

Network connectivity is a problem for this type of device, too. However, it is less apparent than with a notebook, since the amount of data that can be exchanged is much smaller and the speed of entry and retrieval is also much slower so that the network delays play a less significant role.

The great advantage of the phone is that it also contains voice functionality. While this is not the topic of this book, we still need to recognize that the majority of wireless communications for the next few years will be voice based.

PDAs

Handheld personal digital assistants are neither as powerful as a laptop nor as mobile as most mobile phones. Nonetheless they play an important role in bridging the gap between the two. While they do not natively run all desktop applications, there is often an equivalent that will run on many handheld platforms.

They are clearly more mobile than a notebook and offer form factors that approach those offered by very functional phones. They also have a valuable mobile feature of being "always on." It is possible to use them sporadically without needing to restart the system.

In order to provide any type of wireless connectivity it is necessary to either include a wireless PCMCIA card or connect the PDA to a mobile phone—for example, using Infrared. In contrast to notebooks not all PDAs support PCMCIA slots. They are also not likely to have a serial port to which a phone can be connected.

It may be possible to use a CompactFlash card instead, but the most common approach will be via Infrared, a technique that is easy to set up but can be awkward to maintain reliably since it does require line of sight.

Matching the device to the access method

As mentioned previously, the first place to look in order to determine what mechanism to use to access an application wirelessly is the set of clients that need to be supported.

As illustrated in Table 5.2, not all approaches will work with all devices.

This is only an approximation of the most typical devices and their configurations. There are, for example, phones that do have a Web browser on them, and they may some day have the power to run some of the other applications too.

Likewise, there is nothing to stop you from putting a WAP simulator on your notebook or from plugging in a GSM card to either the laptop or PDA and running some software that can send and receive SMS messages. However, these are not typical cases.

5.3 Enabling applications for wireless access | 127

Table 5.2 *Typical Devices and Their Configurations*

	Notebook	PDA	Phone
Native	✔	✔	
Enhanced	✔	✔	
Web	✔	✔	
WAP		✔	✔
SMS			✔
Terminal services	✔	✔	
Proprietary	✔	✔	

You might argue that there is no reason to put a WAP browser on a PDA, and it is certainly unlikely to be a common practice. However, you must keep in mind that the WAP specification is evolving to provide a much richer set of UI functions. At the same time it is much more efficient with bandwidth and can offer functionality similar to the Web but deliver it much faster.

It is possible to develop proprietary solutions on any open platform. This excludes most phones (except those now appearing based on Symbian) but presents an opportunity for notebooks and PDAs.

Additionally, it is possible for the hardware vendor of a proprietary platform to also offer a tailored solution. This means that the phone manufacturers could develop unique solutions that are optimized for their platform. The RIM Blackberry is also an example of a dedicated device that comes with a small set of tailored client solutions.

But client devices are not the only factor for consideration. In Table 5.3 we also see a rating of bandwidth requirements, richness of functionality and user interface, and the provision of local storage.

Clearly the most demanding implementation in terms of bandwidth is a native interface that has not been designed for slow and unreliable networks. Depending on the nature of the wireless networks that are available, this may be completely out of the question. On the other hand, if it is possible to use a wireless LAN or 2.5G network the performance may still be within the realm of acceptability.

Table 5.3 *Other Trade-offs to Consider*

	Bandwidth	Functionality	Local Storage	Effort of Installation	Alerts
Enhanced (Infowave)	High	High	Extensive	Medium	Yes
Web	Medium	High	None	Minimal	No
WAP	Low	Low	None	None	No
SMS	Low	Very Low	Minimal	None	Yes
Terminal Services	Medium	High	None	Minimal	Yes
Proprietary (RIM)	Medium	Medium	Limited	None	Yes

There are other options that require less bandwidth, but you will note that they also offer less functionality. The trade-off is obvious, since any meaningful information needs to be transmitted and will occupy bandwidth. The only way to improve this ratio is by compressing the transmission, an approach used by Infowave.

Local storage is an important factor if the device will sometimes be in use without network connectivity. This could be because the user is in a remote area and does not have access to any network in the vicinity. Or it could also be because the network charges are prohibitive and the user would simply like to peruse some old messages or draft replies to them. In either case it is necessary to operate on off-line storage, which can subsequently be synchronized with the server.

The configuration and support of mobile devices can be a monumental effort. There is great value in being able to use applications that already come preinstalled on the device or that are typically already installed for wireline use.

Alerts can also be very important, since they provide a mechanism to push information to the user. Pull technologies are sufficient for many applications; however, they rely on the user knowing when to browse for more information. If new information arrives at random intervals, the task of polling can be frustrating to the user.

As you look over the trade-offs and the clients that you need to support, you may very well find that there is no single solution that will accommodate all your requirements. Some users, for example, may want only very limited information, but will want it delivered to them immediately to the device they are most likely to have with them: Usually the phone. On the other hand, others may find the machine interface to a PDA and/or notebook much easier to use and will not demand immediate notification.

The environments in which users may need to access their applications are also likely to be different, not only between users, but even for the same users at different points in time. Your system manager may want to be able to monitor the system performance on his or her PDA when in a business meeting. But when eating at a restaurant on the weekend, SMS notification that the system crashed might be more reliable. And a WAP interface to reboot the system may even let a user finish dinner without prolonging system downtime.

A truly complete solution will usually involve several, and conceivably even all, of the approaches described here, if it is to satisfy all the needs of a varied set of users.

Selection of wireless applications

Criteria for initial applications

Any application can potentially be used wirelessly. But some applications will have more difficulty with the unreliable networks and the restricted machine interfaces of typical mobile devices than others. Probably the best criteria for selecting the suitability of an application are its data volume and the value of the information it contains. Clearly the lower the total volume, the faster it will perform and the easier it will be for the user to enter and read. And the user will be most likely to make the additional effort if he or she perceives the content to be of high value. A high value:volume ratio summarizes this trade-off, if not numerically then at least conceptually.

Market segmentation

One obvious division in the wireless market is a split between corporate and consumer segments. Corporate employees will typically also have Internet access through a desktop system but will be interested in wireless technology as a means of attaining mobility. Consumers will not necessarily have access to a PC. For many, the wireless device may be their only, or at least primary, means of connecting to the Internet.

The distinction between corporate users and consumers is, however, not always a simple one to make, particularly in a mobile environment. Many employees already work from home using their own computers and phones or take advantage of their companies' equipment to handle personal matters. In a mobile scenario the devices move from work to home, as well as

Table 5.4 *Correlation between Consumer and Corporate Applications*

Consumer	Corporate
E-mail	E-mail
Instant messaging	Instant messaging
Public information (news, weather, horoscopes, traffic conditions)	Knowledge management portals
Games	Address book (phone number, address)
Ticketing/reservations	Time management (schedule, action items, appointments)
Driving directions	Enterprise resource planning
On-line purchasing	Customer relationship management
Banking	Sales force automation
Stock trading	Supply chain
Auctions	Custom databases
Betting	

many public places on the way. In spite of the efforts of all accountants to control potential misuse, the fine line is likely to blur.

As shown in Table 5.4, there is a correlation between the types of applications that are likely to be deployed and the target market. In particular the corporate market is likely to focus on the same applications they offer on their private intranet. Those targeting home users will probably offer a wider variety of services, which may mirror those found on the Internet.

There is some overlap in the area of messaging (both instant messaging and traditional e-mail), which spans both segments. We also need to keep in mind that there is no clear demarcation in the two user populations, since all employees are also potential consumers of public services. The main implication of the division is in who would offer the services (the employer or a public service provider).

5.4 Consumer applications

As we all know, it is hard to predict which technologies the consumer market will embrace. Stock quotes and news have been widely publicized as areas with high consumer appeal, but they only cater to a portion of the public. Some of the most popular wireless applications to date (based on SMS and Japan's i-mode) have been horoscopes, chatting (instant messaging), and dating services.

5.4.1 Mobile commerce

There is substantial potential for commercial transactions using wireless devices. The built-in security (caller identification provides a basic level of authentication) makes them more suitable than desktop systems anonymously connected to the Internet. In an attempt to tap a new market, some Internet storefronts have already begun to offer the ability to wirelessly order products.

To date, this channel has not yet been widely adopted by consumers, mainly because they still prefer to conduct business using a richer user interface. It remains to be seen if this is a question of merely needing to adapt to a new technology.

The advantage to the consumer is not obvious except in cases where the need is immediate and can be fulfilled instantaneously. Often cited examples are buying drinks from a vending machine or purchasing tickets at a theater. For small amounts of money the device facilitates the transition to a cashless society. For larger transactions it represents the wish of the mobile operators to enter the financial sector, competing with the role of the credit card.

Storefronts may prefer wireless commerce over Internet transactions because of its better security. Credit-card fraud is a significant problem for current Internet Web sites. If mobile commerce is able to establish itself, there would be a strong financial incentive for storefronts to favor it over unauthenticated credit-card transactions.

Ultimately the success of m-Commerce will depend on how widely it is adopted by consumers. While their interest is only marginal, both storefronts and wireless carriers have strong reasons to promote it. It remains to be seen whether they can make the offering appealing enough to their customers.

5.4.2 Mobile Banking

There are many opportunities for financial transactions to be extended to mobile users, including the following:

1. Balance-sheet details: Cash and transaction history
2. Cash transfers

3. Bill payments
4. Portfolio management
5. Personalized stock watch
6. Market stock charts/rates/market movers
7. Trade: Buy and sell stocks and mutual funds

There are many similarities between mobile commerce and mobile banking. In both cases the incentive is related to the increased security provided by an authenticated connection. In some cases, this is further secured by the use of a second SIM card in the device.

Again the mobile operator is in favor of mobile banking transaction, if for no other reason than the air-time charges. The financial institution also benefits from secure transactions.

One major difference is the advantage to the consumer. Fraud on the Internet is mainly a burden for the stores. On the other hand, unauthorized access to bank accounts could ruin the financial situation of an account holder.

Some banks have issued hardware devices (often smart card and smart card readers) that provide a high degree of security to Internet banking. However, the effort to install and use these is greater than it is to use a phone, which already has these built in.

5.4.3 Advertising

Wireless advertising is a controversial issue. There may be a place for it; however, it is important to ensure that the content is highly relevant and not overly intrusive. In contrast to typical Internet browsers, WAP browsers do not have sufficient display area to respond to a user request while simultaneously including unsolicited commercial information.

5.5 Enterprise applications

The main reason for corporate wireless applications is to allow users to continue working when they are not at their desk. In some cases, this may be while the employee is traveling, at another facility, or at a customer site.

Beyond normal work hours, mobile technology also allows users to briefly interrupt vacations and weekends in order to make any necessary decisions or grant approvals, permitting business to continue. While the relative merits of this are controversial the benefit to business of increased productivity can cost justify investment in wireless technology.

5.5.1 Calendar/contacts

Calendar and contacts are often provided with mail applications but can also be offered independently. Both of these involve a small amount of data and are very important to users, even when they are away from their home offices.

In addition to personal calendaring and contact information, users may make ad hoc queries to their colleagues' calendars or may need to access the corporate address lists on their mail server.

Additionally, since many devices have their own Personal Information Manager, one potential application can be the synchronization of appointments and business cards with the corporate server.

5.5.2 Intranet and knowledge management portals

Most corporations have collected a wide variety of partially structured information on their intranets. The degree to which this is useful for mobile employees varies greatly. Given the size of the intranet in most large companies it is unlikely that it will be redesigned for wireless access in bulk. It is more likely that Web sites will be converted as they are needed.

Knowledge management portals, such as those offered by Plumtree and Autonomy, offer users a way to optimize their access to the Internet by creating personalized views of data from a variety of sources. Wireless extensions to these portals help to assist mobile users to navigate to the desired information efficiently and effectively.

5.5.3 Database applications

Databases are among the most common pieces of software being used in business environments, so it is natural to try to port some of this functionality to some of the mobile devices. The challenge is to provide a fully functional database without monopolizing all the resources of the smaller systems.

There are a few solutions that have been able to provide very powerful database capability on some of the mobile platform. They are as follows:

- Oracle Lite (Palm and Windows CE)

- SQL Server Windows CE edition (Windows CE only)
- HanDBase (Palm and Windows CE)

Oracle Lite

Oracle Lite is an Oracle8-compatible object-relational database designed to be embedded inside distributed client applications. It also runs on Windows CE with a 1 MB footprint and supports replication. In line with Oracle's programming methodology it also provides extensive Java support.

The Oracle Lite Consolidator provides a replication mechanism for Palm applications and data to be replicated, synchronized, and shared with the Oracle8 server using the standard HotSync process.

SQL Server Windows CE Edition

SQL Server CE extends the SQL Server database to the CE platform by using a small memory footprint of approximately 1 MB. It exposes a programming and operational model consistent with the rest of the SQL Server family. This means it utilizes the same programming interfaces, including ADO and OLEDB, on the device, as well as being programmable with the eMbedded Visual Tools.

Encryption of 128-bits ensures local database file security. In order to synchronize the data with the server, it has built-in capabilities of remote data access and merge replication using HTTP. It also supports encryption.

HanDBase

At the lower end of the price spectrum, another database running on both Palm and Windows CE is HanDBase. Similar to Microsoft and Oracle it provides server synchronization as well as limited development and import/export capabilities.

5.5.4 CRM applications

Databases alone are not a compelling proposition for the typical user. Unless there is an application running on top of the data they are not of great value. While many custom applications have been and continue to be developed, one specific area that has received a lot of attention along with the e-business revolution is customer-relationship management (CRM).

As with all new terms, it means different things to different people, but its main focus is on helping companies acquire, retain, and serve their customers. To this end the account managers and customer-support representatives need to have a wide variety of customer- and product-related data, including the following:

- Customer contact and profile information
- Logged problem reports
- Order status and tracking information
- Product lines and availability

One of the market leaders in CRM is Siebel. While parts of the Siebel application are available via WAP, the full Siebel application runs on the Windows platform. In order to access it from a PDA one option would be to use Terminal Services.

Sales force automation

The sales force is typically one of the most mobile segments of the employee population. If their applications can be wireless enabled, it reduces their need to return to the office or find wired connections, thereby making them more productive.

5.5.5 Electronic mail

The most prevalent general-purpose corporate application is electronic mail. It also represents information that very often is time critical and therefore lends itself to wireless technology. The disadvantage of mail is the volume of information (number and size of messages) it entails. In order to be used effectively with small devices it requires either good filtering of mail sent to the device or else a dedicated mailbox used only for specific reasons.

A wireless electronic mail solution will typically require a mail application, such as Microsoft Exchange or Lotus Notes, as well as some additional software that is able to provide access to the mailbox wirelessly. Some of these portal applications use industry standards, such as POP, IMAP, and SMTP, and can therefore interface to most modern mail applications. Others use proprietary interfaces, such as MAPI for Exchange, which may provide more functionality but will not interoperate with other products.

Is e-mail a good match for wireless?

You might argue that it is not. After all it is clearly a high-volume application, where daily quotas of several megabytes are a common occurrence. These data also have a fairly low average value. While some of our most important information is sent to us via mail, it is outweighed by the masses of junkmail that flood our inboxes.

At the same time e-mail has the advantage that it is used by almost everyone with an Internet connection. It is the only personal application with this wide a reach. There are many product offerings for wireless access to e-mail and, given the increased rate of standardization of mail protocols and formats, the solutions also are often interoperable.

E-mail also provides a good illustration of how the mode of operation may need to change in a mobile scenario. It is unlikely that the user will want to process all mail on a mobile device. Instead, wireless use will be restricted to quick access, alerts, notification, screening, previewing, and other forms that filter the content down to a scale that is suitable for the device.

Microsoft Exchange

We will discuss some of the mechanisms mentioned previously with a particular application, namely Microsoft Exchange.

Native interface to Exchange

The native interface to Exchange is Outlook. Since it is written for Windows platforms, it is not a general solution for all mobile devices. However, where it does run, it is a very compelling solution, mainly because it is a standard solution that is included in Microsoft Office and therefore already installed on many devices (Notebooks) in order to run e-mail using fixed lines. There is no incremental effort involved in making it also work over a wireless link.

However, there is one big disadvantage to this approach. It offers very poor performance over unreliable and slow network links. It is not efficient in its use of bandwidth and is very vulnerable to high latency. In a high-speed environment the inefficiencies are scarcely noticeable, but over slow dial-up or wireless links they can be unacceptable.

Optimized access

It is possible to optimize the native interface so that, transparent to the user, the network connection is both secured and compressed. One implementation of this approach is Infowave for Exchange.

Infowave creates a VPN between the device and an Infowave server on the corporate LAN. This VPN is especially designed for wireless connections. It compresses data to reduce the transmission size and provides more tolerance in the case of link failures. Additionally, it offers the user some functions to further optimize slow and unreliable wireless connection. Some features include the following:

- Batch processing of e-mail send/receive—optimize the slow e-mail protocols
- It can give the user flexible options for handling attachments, such as the following:
  - Don't send
  - Send only specific pages
  - Send only an abstract
  - Convert to ASCII text and send entire file or a percentage of the file
  - Fax the attachment to me, print, or forward

It also can allow the user to begin to read and respond to mail while downloads/uploads are in process.

Web interface

Outlook Web Access offers an HTTP/HTML interface to Exchange. It is very nicely designed and represents a good alternative to the native Outlook interface for many situations. It only requires a Web browser, which is available on almost every device. While it provides the richest user interface on Microsoft platforms, it is able to operate on virtually any browser and operating system.

This means that typically no incremental installation is needed. It also reduces the need for any additional maintenance, since upgrades can be integrated into standard servicing procedures. The end-user support will also only be related to either the wireless connection or the e-mail application and therefore does not imply any additional expertise.

The speed of access is not overwhelming but can certainly be tolerated across most networks, especially if the amount of information to be retrieved/entered is low. It is possible to optimize the Web network connection, similar to the approach described previously, with products such as Infowave for the Net.

One very important feature/drawback of the Web solution is that it does not support any local storage. In some cases this may be an advantage. For example, when more than one user uses the same terminal this reduces the security risk, since no sensitive information will be maintained across sessions unless it is explicitly stored on the file system.

However, mobile units are usually personal devices. Without local storage the user is not able to work in a disconnected mode. The requirement of a connection is a big drawback in a mobile environment unless the access is only sporadic.

SMS interface

An SMS solution is very well suited for mobile phones, particularly in Europe, Africa, and Asia, since almost all GSM phones can both send and receive SMS messages. SMS also has built-in notification ability, which makes it very useful for alerts.

Some of the solutions, which take advantage of SMS, include Xsonic, Fenestrae, and Microsoft Mobile Information Server—Outlook Mobile Access. However, no two SMS solutions are alike. The extent to which they use SMS can differ in several respects, as follows:

- SMS can be used purely as an alert mechanism (e.g., MIS). Or it can be possible to allow the user to make simple command-like requests to the server (e.g., Xsonic and Fenestrae).

- SMS solutions can integrate with the PIM on the device (e.g., Fenestrae) so that calendar appointments and contacts can be automatically inserted into the phone without any user action.

- An SMS solution can take advantage of a direct connection to an operator's SMSC (e.g., MIS and Fenestrae), which is useful for a large volume of messages; or it can rely on one or more SMS modems, which simulate Mobile Originate SMS messages.

Regardless of the tool that is used, the biggest drawback to SMS is the interface. Manual entry of messages and commands is cumbersome and even reading long messages can be awkward. With many users receiving tens (if not hundreds) of messages per day, a wholesale redirection of mail to the phone is not acceptable.

As part of the solution it is critical that the messages be filtered, and often truncated, according to personalized rules. Only the user can define the criteria (e.g., author, subject, addressees, size) that determine whether an alert is necessary. And only the user knows which elements should be included in the alert (e.g., author, size, first three lines of text, distribution list).

It is quite possible to have an alert-only solution. It simply indicates to the user which messages have arrived, and the user can try to find a fixed connection to view any important messages if desired.

However, a complete solution will also provide a means to selectively retrieve more information or the entire message when the user needs it. The solution does not need to be entirely based on SMS, but some pull mechanism is often required. For example, an SMS message might contain URL for the full message, which can be retrieved via WAP or another interactive protocol.

WAP interface

WAP is primarily a pull mechanism. It is also a primitive interface, but it does allow the user to browse through the mailbox and read individual mail messages or parts of them. In terms of retrieving information the interface is better than SMS, since it is menu-based rather than requiring memorized commands.

The performance of WAP is subject to significant variation depending on the bearer it uses. One of the biggest detractions to WAP is that it is most commonly implemented on circuit-switched data. This means that at least the initial request requires the connection of a dial-up circuit, which is very time consuming. If WAP is implemented on top of a packet data network, the response time improves radically.

Some of the WAP solutions include Infowave FirstHand, Xsonic, Fenestrae, and Microsoft Mobile Information Server, as well as many others that are implemented on IMAP and other industry-standard protocols. One factor to consider is that WAP implies not only the implementation of one of these products but also the availability of a secure WAP connection to the corporate network. This is often a major obstacle, as we will see in the next chapter.

What are MIS and MOMM? Mobile Information Server (MIS) is one of the new members of Microsoft's .Net platform. It is a flexible and extensible framework, which will facilitate the delivery of applications to mobile devices. Microsoft calls it a unified platform for building device-independent applications.

There are two editions of MIS: A carrier and an enterprise edition. In the most typical scenario envisaged by the developers these two will be used in a complementary topology. It is, however, also possible to use only one of the components.

The product is intended to be a framework more than a solution. It does ship with one out-of-the-box application: OMA—Outlook Mobile Access. However, it is only a starting point. The purpose of the technology is to enable development of further applications that can be plugged into MIS.

OMA supports both browse and push access. In V1 browse is supported via WML and push via SMS. It uses an event source that must be installed on the Exchange 2000 server in order to support push.

Another Microsoft product that offers mobile e-mail is Microsoft Outlook Mobile Manager (MOMM). It runs on a desktop and uses the idle cycles to process a user's mail. It can run with or without Mobile Information Servers. Its functions include the following:

- Prioritizing—it has a complex rules system, which is capable of learning, and will use a given set of criteria (sender, size, etc.), will weight them, and will allow you to only forward messages that have a specified priority level.

- Intellishrink—in order to compress longer messages so that they will fit in short SMS messages and will consume less of the small mobile displays, you can specify different shrinking levels. These range from compressing white space, to removing spaces, to removing vowels (as in: ThsRngFrmCmprssngWhtSpcToRmvngSpcsToRmvngVwls in the fully shrunken variant).

Any user can install MOMM on his or her desktop. When a user is in a non-MIS configuration, the mail is forwarded transparently via MSN. This is done because, even if the device is SMTP-addressable, the user would not want to see all the forwarding headers on the device. MSN does some device-specific processing to render it in a suitable format. Once an MIS server is installed in the corporation, then MOMM will sense that and will forward the messages to the MIS server rather than to MSN.

Clearly this means that in the MSN variant the user's mail is being auto-forwarded outside the corporation and there is very little the IT department can do about it. This might provide some IT departments with some additional incentive to install MIS.

Synchronized solutions

Not all mobile solutions require wireless connectivity. If the device has local storage, it is possible to read and create mail off-line and then synchronize with the server whenever a connection is possible again. That connection could, of course, be wireless. But it could also be using a dial-up line of high-speed LAN connection. The principle is the same.

One synchronization solution is from Extended System, which offers an XTNDConnect Server that provides an extensive mobile solution, including synchronization and management of data on mobile devices.

Proprietary interface

Since standards for mobile applications are only now taking hold, there have been some trailblazers who have already employed proprietary solutions and have thereby gained a share of the marketplace. They have the advantage that they can offer specialized hardware, which is tailored for e-mail access.

However, the proprietary approach carries some inherent disadvantages. If the approach uses a closed platform, then it is not possible to extend and/or customize the functionality to suit the end-user requirements. This can mean that while the device might be ideal for e-mail access it must be complemented by an additional mechanism for other applications. Or, in other words, yet another device.

Blackberry is a good example of this type of approach. It is a wireless e-mail solution developed by Research in Motion (RIM) that makes uses of packet data networks (Mobitex) to deliver mail to remote Microsoft Exchange users. It is typically always turned on and therefore (since it is based on packet data rather than a virtual circuit) always connected and it supports notification.

While the RIM Blackberry has been very popular in North America, it has not yet been widely deployed elsewhere. One of the main obstacles has been the network that it uses. If and when it is available on GPRS worldwide, it may be able to expand its success to other continents.

One of its features is the inclusion of a miniature keyboard. As mentioned previously, this is a very useful machine interface for composing e-mail. Its size makes it more awkward to use than a full keyboard but certainly much easier than a similarly sized phone keypad, for example.

RIM offers two products, depending on whether it is intended for a single user or a corporation.

BlackBerry Desktop Software installs and runs locally on an individual desktop PC. It is an integrated suite of applications that provides access to the individual's mailbox and provides functions such as e-mail/organizer synchronization, folder management tools, e-mail filtering capabilities, information backup utilities, and an application loader. This version uses the desktop as a relay station and requires that it remain turned on.

BlackBerry Enterprise Server is optional add-on software that is installed on a server. It can provide mailbox access to all users of one Microsoft Exchange site. E-mail redirection and message encryption occurs at the server rather than the desktop. It provides centralized administration, performance monitoring, configurable handheld security attributes and asset tracking tools.

Integrated access

One key point to note in the previous discussion is that each of the solutions has its advantages and disadvantages and a different set of devices and environments to which it is suited. In fact it is questionable whether any one approach, on its own, can satisfy all the needs of a large user community.

While one or two of the models may be excluded, it is likely that almost all of them will need to be reflected in any complete mobile solution that seeks to provide optimal wireless access to Exchange.

Lotus Domino solutions

Lotus Domino has access to a similar set of wireless solutions as Microsoft Exchange including the following:

- Real-time access to Domino Server
- Web access to Notes databases
- Mobile Notes/Mobile Services for Domino
- RIM BlackBerry support for Notes
- Motorola FLEX Messaging Server
- Extended Systems—synchronization with Notes mail, calendar, databases

Mobile Services for Domino supports a wide range of devices from one-way pagers to PDAs using any of the wireless data and paging networks. With a specially designed micro-browser the users can perform most of the typical mail operations including read, reply, delete, forward, compose, and send mail. They also have access to the powerful search capability of both the directory and the inbox.

Novell GroupWise 5.5 solutions

Novell is no different from Exchange and Domino in having a wide range of mobile support, which includes the following:

- Real-time access to GroupWise
- Novell GroupWise Wireless
- Puma Intellisync Synchronization
- RIM Blackberry support for GroupWise

Novell GroupWise Wireless is included with GroupWise 5.5 Enhancement Pack and supports both WAP and HDML devices with the typical mail functions as well as the ability to initiate a voice call from the address book.

Internet mail protocols

An alternative to the mail system–specific solutions is software that utilizes the Internet Mail protocols. It may be less optimized for many of the actions, and it is often less functional than some of the hooks available to proprietary solutions. But it is also much more transparent in its actions and, above all, it is more interoperable.

In theory, almost all mail solutions can support the Internet protocols. So for users who may need to access different mail systems this may be the only efficient option. Mobile devices are especially likely to be used in different situations, which makes them excellent candidates for these protocols. In fact, many of the solutions, even from mail-system vendors, ship with these out of the box. An example of this would be the PocketPC, which uses IMAP rather than Microsoft's earlier favorite, MAPI.

Within the Internet protocol suite there are several that are related to mail. Interactive Mail Access Protocol (IMAP) is particularly suited for on-line access, since it gives access to a folder structure. Post Office Protocol (POP) is geared toward bulk download, which makes it more appropriate to a synchronization model.

Simple Mail Transfer Protocol (SMTP) can be used to send messages either instantaneously (in the case of a connection) or in bulk (saved until synchronization).

These three protocols use a format called Multipurpose Internet Mail Extensions (MIME). There are other formats that are suitable for nonmail items, such as calendar (vCal) and contact information (vCard).

Potential improvements in e-mail

The previous discussion explains different methods of accessing mail in an efficient manner. What it doesn't cover is how a mail system can improve its use for mobile users. There are a number of ways it would be possible to reduce the amount of unnecessary volume and therefore increase the average value of the information. These include the following:

- Improved filtering and prioritization of messages can use more complex rules and heuristic techniques to learn based on feedback from the user.

- Compression provides a way to transfer less data. For example, the application can reduce excess white space or even remove it altogether (e.g., ThisIsACompresssedMessage).
- Personalization provides a mechanism to define shortcuts.

Unified messaging also may play an important role in the domain of wireless. Many mobile devices, such as phones, are equipped for voice use. By automatically connecting voice with e-mail it can be possible to reduce the user's need to enter and read long messages.

5.6 Vertical applications

Most of the application mentioned so far are considered horizontal in that they are equally useful for all industry segments. E-mail, for example, is used by financial, manufacturing, and healthcare professionals in the same way. But beyond these cross-industry applications there are also many specific needs that relate only to a very targeted audience. Some of these are listed in Table 5.5.

For all corporate applications the challenge is to create a unified solution that optimizes user access based on a variety of information needs regardless of the type of device and transmission standards used.

Table 5.5 *Specific Application Needs*

Fleet messaging	Package delivery and pickup. Packages can be scanned at each point en route so the system can inform customers where packages are at any time. Delivery may be confirmed with a signature captured on a screen with pen input. System-wide status can be updated within seconds or minutes.
Field service support	Rapid dispatch of repair personnel in response to customer-critical needs. It can facilitate efficient scheduling and rescheduling of on-site visits and immediate access to parts availability, prices, and help from other personnel.
Dispatch	Efficient scheduling and routing. Taxi companies and limousine services can use wireless dispatch applications. Messages can be displayed on dash-mounted units and delivered even when the driver is out of the car.
Healthcare	Real-time access to patient information from any location, inputting of patient details, patient histories, and prescription information and medical supply inventories.
Insurance	Filing and verification of insurance claims.
Manufacturing	Materials can be tracked and manufacturing processes controlled more efficiently.
Retail	Increased efficiency of order taking and tracking, inventory management, and faster customer service

5.7 Optimized wireless applications

Current applications are not really designed for mobile users. They were developed with desktop users in mind. In order to make them available to wireless devices we needed to find innovative solutions that made their use acceptable.

With most projections estimating that mobile devices will soon outnumber fixed devices this approach needs to be revisited. The task might seem impossible given the wide range of networks and devices that are available. But as the infrastructure migrates to IP-based networks, the effort on the part of the application can concentrate on accommodating the needs of the growing spectrum of mobile devices.

If a solid wireless infrastructure is in place, as suggested previously, then the applications do not need to take "wireless" into consideration. However, it is important for applications to be developed with the new model in mind. This has some specific implications, which should be factored into the approach, as follows:

- Resiliency
- Synchronization
- Dynamic presentation

5.7.1 Resiliency

The application and, in particular, the platform on which it resides, must be supportive of slow and unreliable connections and dynamic reconfiguration. For example, the application should not time out in a high-latency situation. It should also be able to recover from errors in transmission. A more challenging requirement for some platforms may also be the ability to respond to dynamic reconfiguration as the devices switch interfaces or networks—and thus receive a new IP address.

5.7.2 Synchronization

One important observation with respect to mobile devices is that they are not always connected. This may be because there is no local coverage, because their use is barred (e.g., hospitals, airplanes), or simply because it is not cost effective to use them at a given point in time. This implies that the device must be usable offline and must be configured with local storage. When it is reconnected, it must be possible to synchronize the changes that have been made. There are several proprietary synchronization mechanisms—and one powerful initiative, SyncML—that seek to standardize a protocol for multi-platform use.

5.7.3 Dynamic presentation

As we mentioned earlier in the chapter, there are two fundamentally different approaches to client/server presentation.

Client presentation

The first implies a loose coupling between the client and the server. All presentation is produced by the client device, which only has one machine interface to worry about. It exchanges application information with the server but no display information. This approach can produce tailored content for the user and is able to optimize the network link. However, it has the drawback that it is not generic and requires a specific program to be installed on the mobile device for every application. If this software has a large footprint, then it may impact the number of applications that the user is able to install.

It also needs to be able to process the data formats being transferred from the server over the chosen transport protocols. Since these data are stored locally, off-line access is possible but implies a need for synchronization.

Server presentation

Another approach is for the server to render the content and to provide pure display information to the client—for example, through HTML or through a terminal window such as Citrix or Microsoft Terminal Services.

In this case, content must be redesigned, preferably on an XML basis, and must be rendered in a device-specific format. One aspect of this is that some devices will interpret WML and other cHTML, HTML, or XHTML. The syntactic difference is but one consideration. The decision of how much and the selection of which content to transmit is more closely related to the form factor. The machine interface will also affect the dialog in a fundamental way and must be considered in the presentation.

No local data need to be stored (although it might be useful to cache recently viewed information), but the user will not be able to actively work when disconnected.

5.8 Increasing the value

The potential of wireless networks is continually increasing. As applications begin to take advantage of this evolution they too will begin to offer new kinds of functionality.

5.8.1 Location-based services

One advantage of mobile devices over PCs is the possibility to determine the user's location and provide location-based data, which provides a means of automatically targeting relevant information.

There are numerous applications that can benefit from this information, including the following:

- Interactive maps and driving directions
- Location-dependent contact information
- Location-dependent travel information
- E-commerce—proximate store/bank/restaurant/gas station locations

Wide area services

The most prominent type of location-based services is potentially global in scale. It offers only limited precision (e.g., 100 m), but this is sufficient for many types of applications. For example, it can give optimized information about flights, as well as addresses, local traffic information, directions and maps of restaurants, pharmacies, cash dispensers, gas stations, parking garages, and hotels, to name just a few.

There are two different approaches that can be used to extract location information of this granularity: GPS and carrier integration.

GPS can provide global coverage independent of the particular air-interface that is being used. The mobile unit must be equipped with a GPS receiver and pass its coordinates to the server application. The server can then supply information that is specific to the location of the user.

Wireless infrastructure providers, such as mobile operators, also have information on the location of a given user at any point in time. They can derive it by comparing the signal they receive from multiple base stations. And they must perform these calculations anyway in order to efficiently transfer the mobile user from one base station to the next.

However, this does not mean that the operator is legally allowed to release location information to a third party. And even if there are no legal obstacles, the operator may charge considerably to release these valuable data. If it is possible to come to an agreement, this approach has the distinct advantage that it does not require any incremental support on the part of the mobile unit. The server application can communicate directly with the carrier so that the user of any device can transparently access location-specific information.

High-precision services

A very different kind of location-based service is useful inside buildings or on a small campus. It cannot easily cover a wide area but can deliver very precise location information. An example of this would be an art museum, where mobile users could receive information on the painting nearest to them as they move about.

It could also provide directions in large office buildings and manufacturing plants, or even public areas such as airports, shopping malls, or amusement parks. Directions and locations of nearby services are always the most obvious application of a location-based solution. The amount of additional content is variable and depends on the context and the preferences of the user. Some examples might include special offers/opportunities (new ride, store, airport lounge) or problem reports (flight canceled, long line at immigration, ride closed).

Unfortunately, there is not yet any standard mechanism to provide this information. There are many different efforts that involve either identifying which access point is being used (e.g., by IP address) or else equipping the terrain with special beacons, which serve the sole purpose of identifying the location of the mobile units.

5.8.2 High-bandwidth services

The arrival of 2.5G and the advent of 3G are complemented by existing short-range broadband technologies, such as wireless LANs. As these comprehensively offer larger coverage areas of high-speed wireless networks, applications can begin to take advantage of them for newer services, including the following:

- Rich media Web browsing
- Video chat and conferencing

- Streaming audio and video
- Real-time multi-player games

5.9 Summary

There are many different approaches that can be used to provide mobile access to applications. One of the most fundamental distinctions is whether the presentation should be produced by the client or the server. Terminal services, Web and WAP browsers use server-side presentation while many of the native solutions use client-side presentation. There are also distinctions in whether the solution must support off-line and/or real-time access to the applications. And, clearly, not all approaches are available for all client platforms or back-end applications.

In order to find a suitable approach for a given environment, it is important to look at the set of client devices, the required applications, and the environments where the device may be used. In many cases it may not be possible to find a single solution that can meet all the requirements, so it may be necessary to find multiple complementary approaches that cover the spectrum of client needs.

Related Web sites

Microsoft Mobile Information Server: http://www.microsoft.com/miserver/

Novell Groupwise: http://www.novell.com/wireless/

Lotus: http://www.lotus.com/wireless

Extended Systems: http://www.extendedsystems.com/

Infowave: http://www.infowave.com

Wireless Knowledge: http://www.wirelessknowledge.com

Research in Motion: http://www.rim.net

Xsonic: http://www.xsonic.com/

Microsoft SQL Server CE: http://www.microsoft.com/sql/CE/

Oracle Lite: http://technet.oracle.com/products/8i_lite/content.html

6
Security

Most industry analysts project that mobile Internet devices will outnumber fixed devices by the end of the decade or possibly even sooner. While the adoption rate to date has been impressive, it has also met with great resistance due to security concerns.

This chapter examines some of the issues related to wireless and mobile security and explores the solutions that are available to address any points of vulnerability.

In many ways wireless security is just like wireline security. The issues are largely the same. Regardless of the medium every system needs to safeguard proper authentication, privacy of transmission, prevention of viruses, and protection against denial-of-service attacks.

The differences occur from the fact that mobile devices and transmission over an unshielded medium (air) are inherently more vulnerable to impersonation, sabotage, and interception.

6.1 Device security

6.1.1 Problems with mobile passwords

Keeping passwords secret is a challenge in any environment. But the nature of mobile devices makes them even more susceptible than fixed terminals. The miniaturization of the computer, which implies small user interfaces and keypads, leads users to select even simpler passwords than they would in the wired world. Since multiple letters are associated with each numeric key on a keypad (differentiated through iterative key presses: e.g., press once for A, press twice for B), many users choose words that use the first (single press) letters, thereby substantially reducing the number of possible passwords.

Many phones have "intelligent" entry, such as iTAP or T9, which will perform dictionary comparisons to make text entry using a phone keypad easier. The user needs to enter fewer key presses per letter. If this software is activated during password entry, the user may be more inclined to choose longer passwords; however, the combinatorial possibilities are still reduced since the keypad uses less keys than a keyboard.

Mobile devices are used more frequently in public, making the chance of shoulder surfing more real. While you have some control over who stands behind you in your own office, it is virtually impossible to prevent someone from standing behind you (often without your knowledge) in public places, such as on airplanes, in lounges, and on trains.

Many users cache their passwords on their device, in order to automate their connection to a server. Stolen and lost machines then provide preconfigured access to the corporate network and applications.

6.1.2 Vulnerable file system

Misplaced and misappropriated devices also represent another security threat. If they contain sensitive data (such as passwords and account numbers or classified and/or confidential information), it is usually not difficult to obtain these data once the device changes hands.

Even password-protected devices often have some means of bypassing security, just in case the user forgets the password. For similar reasons, they usually do not lock out the user when confronted with a brute-force attack.

It is important to verify what happens to data and security if the device is powered down and/or completely reset. While some desktop operating systems, such as Windows 2000, offer an encrypted file system, this is not yet common on most mobile platforms. This opens the door to physical extraction of data if the motivation is high enough to justify the effort.

6.1.3 Smart cards

A partial solution to the problems of mobile devices is the use of smart cards. These hardware tokens provide secure storage of PINs/passwords, private keys, and certificates. They often also include tamper-proof implementations of the cryptographic algorithms themselves. In order to fit on the card these algorithms are usually based on elliptic curve cryptosystems and therefore are less processor memory intensive than the otherwise more common RSA algorithms.

Obviously, these could just as easily be stored on the device itself. What makes these unique is that the tokens are virtually tamper proof and can only be used after authentication to the card.

That is not to say that it is impossible to crack a smart card. Some of the attempts have included modifying the circuits, erasing Electrically Erasable Programmable Read Only Memory (EEPROM) (e.g., by manipulating voltage levels), and attacking the random number generator, which influences the generation of the cryptographic keys.

What is important is that all these attacks operate at a physical level. This means that they require physical possession, sophisticated tools, and advanced expertise, which includes a comprehensive understanding of the physical layout of the card.

A highly recommended book, *Handbuch der Chipkarten*, by Wolfgang Rankl and Wolfgang Effing (Hanser Verlag, 1999), presents an in-depth look at smart cards.

Smart card standards

There are several different types of standards related to smart cards. In particular, there is the physical interface to the card. While this interface would conceivably suffice, it is not practical for applications to operate at this level. In the same way that networks are divided into layers for ease of use, you can imagine the application programming interface positioned above the physical interface.

Most of the physical interfaces to Integrated Circuit (IC) cards are developed collaboratively by International Organization for Standardization (ISO) and International Electrotechnical Commission (IEC). The most common of the physical interfaces is ISO/IEC 7816. It defines the card size; placement, shape, and size of the contacts; the purpose of each of the contacts; the voltages and signals that are applied to the clock and I/O ports; and the command set recognized by smart cards. In a nutshell it covers the whole protocol that goes over these contacts.

The other physical interfaces do not assume physical contact but instead use optical interfaces (ISO/IEC 11694) or wireless (radio-frequency) interfaces (ISO/IEC 10536, 14443, 15693) covering varying ranges.

Public Key Cryptography Standard (PKCS) #11—also known as Cryptographic Token Interface (Cryptoki)—is the de facto most common API to cryptographic function. It was developed by RSA Security and is available across most platforms.

There are two other common application programming interfaces to smart cards. The Personal Computer/Smart Card (PC/SC) interface was developed by a group composed of HP, IBM, Sun, and Schlumberger, as well as several others. Microsoft drove the specification and continues to implement it on most Windows platforms. It has the advantage of being more functional than PKCS #11 but is not as portable.

The OpenCard Framework (OCF) is a more Java-oriented approach with many similarities to PC/SC, including both the advantages and the disadvantages.

It is important to distinguish both OCF and PC/SC from JavaCard and Windows for Smartcards. The latter two are operating systems that run on a card, whereas the OCF and PC/SC run on a computer and serve to abstract the smart card communication details from the applications.

RSA also has another related standard: PKCS #15, the Cryptographic Token Information Syntax Standard. This syntax defines the format of cryptographic tokens (e.g., secret or private keys, authentication objects) that are stored on an IC card. This is regardless of the cryptographic interface used to transfer them.

SIM cards

Subscriber Identity Modules (SIMs) are also a form of smart card. They have become popular with the GSM system, where they are a core part of the specification. The SIM cards authenticate a user to a mobile operator and provide functions for securing the transmission.

GSM SIM cards also use the ISO/IEC 7816 standard. However, they do not use a "standard" application programming interface, since there was no intention for the SIM cards to be used on a grand scale by application pro-

Table 6.1 *Smart card–Based Phone Identity Modules*

Abbreviation	Full name	Defining Standard
SIM	Subscriber Identity Module	GSM
WIM	Wireless Identity Module	GSM/WAP
UIM	User Identity Module	IS-95
R-UIM	Removable UIM	IS-95
USIM	Universal SIM	UMTS

grammers. There is a SIM Toolkit, which can be used to exploit the additional capacity of the cards.

Cards with SIM Toolkit applications can monitor the phone's keyboard and dynamically insert menus. These menu items then trigger applications, which can request information from the end user (e.g., a credit card account number or dollar amount).

As shown in Table 6.1, GSM was the first but is not be the only mobile phone system to use smart cards. Other systems include the following:

- CDMA (IS-95) R-UIMs have been tested in Asia and will be appearing on the market soon.

- UMTS has already specified the use of the USIM (universal SIM) as the evolution of the SIM.

- The GSM ANSI-136 Interoperability Team (GAIT) has finalized the specifications of a multi-mode phone supporting GSM 900/1800/1900, TDMA (IS-136) 800/1900, and AMPS. The GAIT phone also relies on the SIM card for the storage of all GSM and IS-136 authentication, roaming, and service information.

Note that the SIM cards offer additional advantages to the user besides enhanced security. They decouple the handset from the mobile subscription, which facilitates global roaming, since there is no longer the requirement to use the same air interface. According to projections from the Gartner Group ("Mobile Value-Added Services and Smart Cards," July 2000) more than 80 percent of all mobile terminals shipped will contain a smart card (SIM, R-UIM, or UIM) by 2004.

6.1.4 Virus protection

Not all mobile configurations are vulnerable to virus attacks. In order for a virus to install itself it must first be transferred to the device. This implies connectivity. Then it must execute on the device, which implies the ability to run ad hoc applications. The damage a virus can do depends on the privileges and functions available to these applications. Many can modify local storage and therefore perpetuate themselves on the device. In order to be effectively distributed they need to have some means of propagating themselves to other users. Sending an e-mail message to all the recipients in the address book is the usual approach, but other mechanisms (posting files for HTTP/FTP download) are also possible.

For a sophisticated and open system such as a desktop it is easy to meet those requirements. Laptops and PDAs are already exposed to the same risk whether they use wireless or fixed connections. But most mobile tools are still very limited and therefore less susceptible. However, it would be naive to assume that they are likely to remain immune for the foreseeable future. As they grow in power and open their systems, they, too, will be the target of hostile attacks.

WAP phones have not yet encountered significant problems, mainly because most WML is passive. However WMLscript opens the door to active applications. In particular, the telephony interface, WTAI, is a potential area for misuse that could conceivably include expensive phone charges incurred without the knowledge of the user.

The fundamental problem with virus protection is how to defend the system without restricting constructive applications and data. It would be simple to close all the doors, but that would also reduce the usefulness of the machine.

The most common approach is to scan through any executable code looking for a known set of viruses. The biggest question is where to run this search. Enterprises frequently scan at the firewall or on mail servers. But mobile devices are often connected to the Internet without using the corporate network. So it is difficult to ensure that they are not infected.

The solution for laptops may be to run some virus protection software locally. Smaller devices, however, often lack the computational resources to be able to do this efficiently, although there has been some progress, at least on the PDAs. F-secure, for example, has an antivirus solution for the iPAQ Pocket PC.

Even local solutions have a problem with mobility though. New viruses appear regularly, and only a machine that has the properly updated solution is optimally protected. It is therefore necessary to ensure that the newest signature files from the antivirus vendor are redistributed to all devices frequently.

6.2 Air security

6.2.1 Bluetooth

Bluetooth is intended for personal area networks, as a means of connectivity between personal devices and peripherals. Its short range functions as a means of protection against large-scale anonymous attacks. While that

6.2 Air security

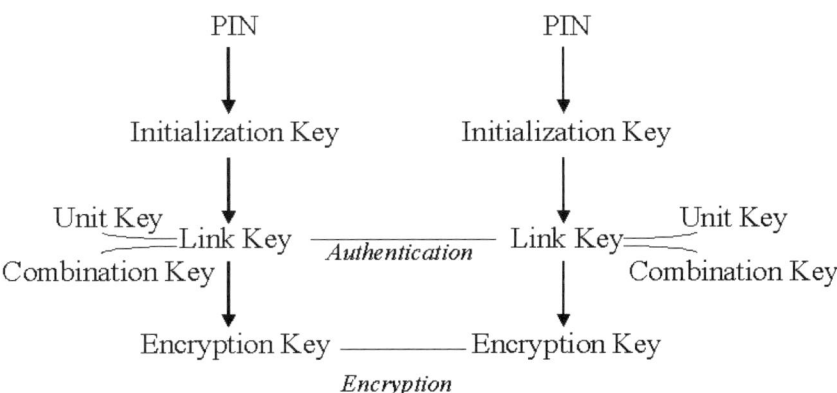

Figure 6.1 *Bluetooth authentication.*

might suffice for the home and other controlled environments, it is clearly inadequate for public environments, where competitors and indiscriminate hackers might also be present.

The ad hoc nature of Bluetooth prevents its integration within a PKI or its use of a Key Distribution Center, that would allow a more sophisticated authentication scheme such as Kerberos. It scheme is confined to the security that can be provided in a peer-to-peer model.

In order to address the need for additional security Bluetooth devices can require authentication before establishing a link. A link key is then established between the devices and all communication can be encrypted. (See Figure 6.1.)

The initial authentication makes use of a PIN, which must be identical on any paired devices in order to set up the link key. Ideally all PINs are configurable on the device, but Bluetooth is also available on peripherals that do not have a UI and must therefore utilize a hard-coded PIN.

Bluetooth security includes the concept of authorization. Each device can be configured with a set of trusted devices, which have unrestricted access to all services provided. Untrusted devices, on the other hand, have access only to services that have not been restricted.

It is beyond the scope of this book to analyze the Bluetooth cryptographic algorithms. However, for the sake of familiarization, some of the more common functions are as follows:

- E0—Encryption
- E1—Authentication
- E21—Unit key generation

- E22—Initialization key generation
- E3—Encryption key generation

Two potential weaknesses of Bluetooth security include the PIN attack and the location attack.

PIN attack

Whether the PIN is hard coded or configurable, it is important to realize that short (e.g., four-digit) PINs have only a small set of possible values. We compound this problem with the fact that many users tend to leave settings with their default values. Consequently, the chances of breaking a PIN—for example, through a brute-force attack, can be significant in a poorly configured environment.

Location attack

It is possible to recognize a device without authenticating it. This is not ordinarily a big risk. However, it is feasible that Bluetooth probes could be installed in prominent locations, which might include airports, entrances to corporate office buildings or even private homes of selected targets. By correlating the traffic it is then possible to identify patterns of movement that may provide insight into business dealings and other discretionary information.

6.2.2 WLAN

Wireless LANs are most often used in corporate environments, where all employees are presumed to have unrestricted access to the network. However, close proximity is not a sufficient factor for authorization, since there may be guests or neighboring offices that share the same air space but should not be allowed to access network resources.

The only way to restrict this today is through the use of Wired Equivalent Privacy (WEP), which can be configured at the access point. Another network authentication scheme, called 802.1x, is currently also being proposed that could augment WEP, if applied to wireless LANs.

WEP

WEP uses a symmetrical algorithm called RC4 with either a 40-bit or 128-bit key. When WEP is enabled, each station (clients and access points) has up to four keys. It is able to provide both authentication and encryption of all data transmitted over the air. Although the algorithm will permit different keys for each user, there is currently no implementation that is able to manage per-user keys. Instead, all devices using a single access point typically share one key.

It should be noted that WEP is only available when an access point is in use. Stations that have formed an ad hoc network cannot enforce authentication and must pass all data in the clear.

There are two problems with WEP. The first is the cryptological weakness in the algorithm that makes it susceptible to statistical analysis. While there have not yet been reports of any actual intrusions staged on a purely cryptological basis, they may be imminent, since they do not require substantial equipment.

Many of these were documented in a report produced by the University of California, Berkeley. To address these and strengthen the security beyond its initial requirements IEEE 802.11 Task Group E (IEEE 802.11e) is currently working on extensions to WEP for incorporation in a future version of the standard.

Another problem is key distribution. Since all users of one access point must share the same password, and most corporations would like hand-over between access points to be transparent to the user, it implies that all users must use the password. This poses a problem if an employee resigns or is terminated. Theoretically, the password should then be changed, but it is not always an easy task to communicate the new password to the entire user base in a timely manner. The larger and more dynamic the employee population the more complicated the process becomes.

As a consequence of these issues, the next version of 802.11 (802.11a) should include security enhancements that address the weaknesses. One part of this is 802.1x, which can be included in any access point and will permit authentication to any authentication database. Per-user authentication eliminates the key-distribution problem.

802.1x

802.1x is not limited to wireless networks. It can be used to authenticate user access to any closed network. For example, a company may have a private network, which should be accessible only to employees, with some public segments, which can also be made available to customers. Without 802.1x it would be necessary to isolate these two networks, which could lead to significant duplication of effort and equipment.

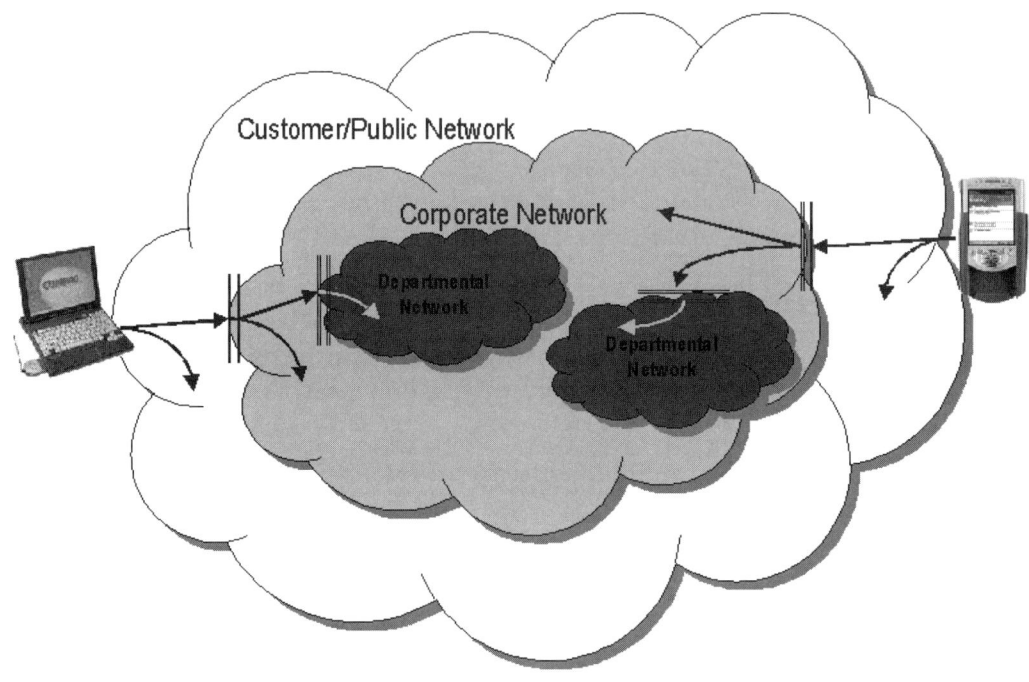

Figure 6.2 *802.1x authentication.*

The fundamental approach used by 802.1x is to authenticate users at the edge of the private network. It would be conceivable to perform this processing at other points in the core of the network—for example, using MAC addresses. However, it would be difficult to protect all authenticated end stations from unauthenticated stations, since intruders could bypass authentication, at least on their own segments. It is significantly less complex, and more scalable, to ensure security if the authentication is performed on the external boundary of the network. (See Figure 6.2.)

It is possible to develop a tiered authentication scheme in which the public is able to access the external network. All employees can access the corporate network, and individuals can access their restricted departmental LANs.

Controlled ports

A typical bridge would connect segments that are private and already presumed to be secure, such as those on the corporate network. An 802.1x bridge can connect these segments, but its added value lies in its ability to optionally authenticate a port before allowing it to connect. (See Figure 6.3.)

6.2 Air security

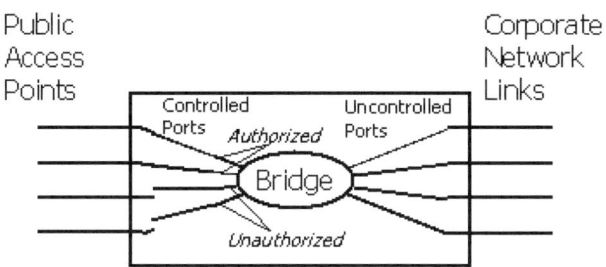

Figure 6.3
802.1x bridge.

What this means is that the bridge is configured with both controlled and uncontrolled ports. Those that are uncontrolled do not need to authenticate and would see the device as though it were a traditional bridge. Devices connecting to the controlled ports would not be able to access any of the connected segments (neither the segments on the uncontrolled ports nor the segments on authenticated controlled ports) until they authenticated successfully.

802.1x architecture

The scenario sounds simple in principle. Where it becomes slightly more complicated is in the actual authentication. Conceptually it would be feasible to let the bridge perform the authentication using a cache of authentication information. However, that would be unnecessary overhead for the bridge and would mean that authentication information would need to be replicated to all bridges, which is neither efficient nor secure.

Instead, the bridge (called the Authenticator) may relay authentication requests from a client (called the Supplicant) to an Authentication Server. This is very similar to the RADIUS model of authentication, and, in fact, it is expected that many Authenticators will be RADIUS clients—and many Authentication servers will therefore be RADIUS servers.

There are three players in this topology. The Authenticator sits in the middle with both controlled and uncontrolled ports. The Authentication server is connected to an uncontrolled port. The Supplicant is connected to a controlled port. (See Figure 6.4.)

The authentication process runs along the following lines:

1. The Supplicant connects to a controlled port.
2. Either the Authenticator or the Supplicant initiates authentication.
3. A challenge is sent from the Authentication Server to the Supplicant via the Authenticator.

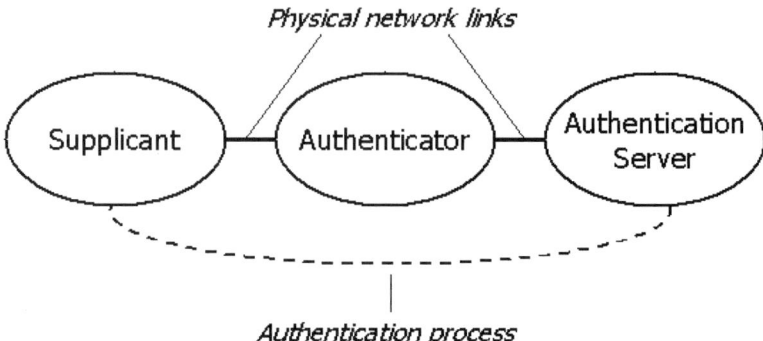

Figure 6.4
The authentication process.

4. The Supplicant signs, or otherwise cryptologically processes, the challenge.
5. The Supplicant sends the result back to the Authentication Server via the Authenticator.
6. The Authentication Server sends the status (success or failure) back to the Supplicant via the Authenticator.
7. The Authenticator intercepts the status and, if successful, opens the port.

Consider the following issues in this process:

- The only traffic that the Authenticator may relay from/to a controlled port is authentication requests/responses.
- For security reasons, the authentication information must be cryptologically secure. This implies that the Authenticator cannot decrypt the credentials.
- The model must be extensible to new authentication mechanisms as they are invented and implemented.

In order to ensure that the Authenticator can always identify and interpret new authentication mechanisms, any authentication types must be encapsulated using the Extensible Authentication Protocol (EAP) as specified in RFC 2284. EAP already supports multiple authentication schemes including smart cards, Kerberos, public-key encryptions, one-time passwords, and others. (See Figure 6.5.)

Security considerations

The biggest security consideration of 802.1x is that its sole purpose is authentication. It does not provide integrity, encryption, replay protection,

Figure 6.5
Extensible Authentication Protocol.

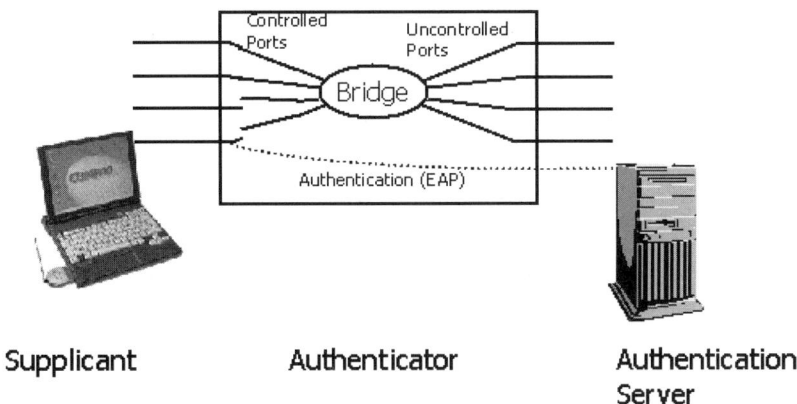

or nonrepudiation. These would need to be implemented with complementary schemes such as IPsec.

There are other points of vulnerability that must be addressed in any implementation of 802.1x. These include the following:

- Piggybacking on an authenticated port—multiple end stations on a port must be detected and disconnected.

- Interception of credentials—passwords must always be encrypted.

- Subversion of authentication negotiation—it should not be possible to provoke a lesser form of authentication by interfering with the authentication process.

Wireless implementation

802.11b wireless LANs are ideal candidates for 802.1x authentication, since they represent a completely uncontrolled periphery. While it is possible to restrict physical access to wired LANs, this is not feasible in a wireless environment. It is much more difficult to monitor and enforce the air space around office buildings than the ports and wiring within them.

This vulnerability is currently addressed using Wired Equivalent Privacy (WEP), which is available on 802.11b access points. If in use, then all stations must configure a symmetric passphrase in order to connect. All transmission is then encrypted with 40–128-bit encryption.

Recently, there have been alleged cryptological weaknesses with the WEP algorithms that have cast a shadow on its use. Beyond these there is a fundamental problem with key distribution and update. Since WEP keys are typically symmetrical (the same on the access point and all connecting stations), they need to be changed in unison. Clearly this is difficult to orchestrate when large user populations are involved.

There have been solutions, including automating regular key changes, for example, using log-on scripts; however, they are nonstandard and require additional work. There are also problems ensuring that employees who leave the company no longer have access to the network, since they could "remember" their WEP key.

Another aspect of the problem occurs when users connect to multiple wireless LANs (e.g., in public areas or at customer sites). Current WEP implementations require that the user manually change the WEP key each time a new network is selected, which is tedious and interferes with any automated key changes.

802.1x solves all of these problems. It is not necessary to distribute any keys. The user can authenticate to a central Authentication Server, which stores per-user credentials that can be disabled or modified as needed.

This does not mean that there is no longer a need for WEP in an 802.11b LAN. As mentioned previously, 802.1x only provides authentication. It does not encrypt the over-the-air transmission. It is, therefore, still possible for hackers to eavesdrop on conversations and intercept sensitive information.

The ideal combination is to use 802.1x for authentication to the network and WEP to ensure privacy of the transmission. This does not address the cryptological weaknesses of WEP; however, it does open the door for future versions of WEP to focus on privacy rather than authentication.

6.2.3 WWAN

Digitized mobile telephone and wireless packet data networks all include some kind of encryption. First-generation (analog) systems are not suitable for data transmission, so the risk of intrusion was limited to eavesdropping on private conversations and no encryption was deemed necessary. Since the migration to the second generation of wireless telephony is virtually complete, we can restrict this discussion to digital networks.

The world's most widely used wireless phone system is GSM, which uses a smart card that contains both the International Mobile Subscriber Identity (IMSI) and the subscriber identification key. Upon establishing a connection with a mobile base station a session key is negotiated and all transmissions, voice and data, are encrypted and are very difficult to crack.

GSM documents specify the rough functional characteristics of its protocols including the secure encryption of transmitted digital messages. However, apart from the protocols, details of the algorithms are kept secret. Most security specialists will argue that security by obscurity is not an effective approach, since only the close scrutiny of a large set of experts can ensure that there are no obvious weaknesses in the technique. Nonetheless, GSM contains three secret algorithms, which are only given to experts with established need-to-know, such as carriers and handset manufacturers. These three algorithms are as follows:

1. A3: Authentication algorithm

2. A5: Ciphering/deciphering algorithm (presently A5/1, A5/2)—provides over-the-air voice privacy

3. A8: Cipher key generator (essentially a one-way function) and (session key generation)

The SIM card contains A3, A5, and A8; the base station is equipped with A5 encryption and is connected with an authentication center using A3 and A8 algorithms to authenticate the mobile participant and generate a session key.

Most of the other wireless wide area networks (e.g., IS-95 CDMA, IS-136 TDMA) were developed in the United States and use the CMEA encryption standard specified by the Telecommunications Industry Association (TIA), also an effective encryption technique. Additionally, IS-95 CDMA uses a transmission technique called spread spectrum, which was developed by the military with the express intent of making interception more difficult.

Although these encryption algorithms provide an effective barrier against the vast majority of hackers, it is important to realize that they are not uncrackable. Both the CMEA and GSM algorithms are reported to have been cracked. The value of the protection does not lie in providing a completely secure environment for very sensitive transactions. Instead, it offers an obstacle so that monitoring and interception of random or bulk transmissions are simply not cost effective.

Network security

In addition to the security of the air interface of a wireless WAN we also need to consider the network between the base station and the application server. Fundamentally there are two different methods a mobile network may offer to transfer data. It can provide a packet-data network or it can use circuit-switched connections.

A packet-data network is simpler. CDPD, Mobitex, and GPRS would all be examples of packet-data networks. In these cases the mobile device has an IP address and it transfers data through the mobile network, which is connected to the Internet. No special configuration is typically required at the mobile end. Its data access is transparent. If the IP address given to the device is fixed, then a minimal amount of authentication is also implicit in any packets originating from it.

Data communication over primarily voice networks, such as GSM, IS-136, and IS-95, is not quite as straightforward. Typically a PPP connection must first be made from the device to a dial-in server. The dial-in server will assign an IP address and relay all the traffic between the device and any application servers.

This implies some configuration at the mobile end. The phone number must be specified, and then the user must authenticate to the dial-in server using an authentication protocol such as PAP, CHAP, or MS-CHAP. So the dial-in server knows who the user is but the application server does not. It cannot determine the phone number easily and the IP address is meaningless. If necessary, it would then reauthenticate the user, which means additional work for the user.

It would be possible to bypass the first authentication by storing the mobile phone number on the dial-in server and then comparing the caller-ID of incoming calls. However, this would provide unlimited access to the corporate network when a device was lost or stolen.

Solutions to address this dilemma must combine security with ease of use—for example, by using biometric authentication. They must also ensure that unauthenticated users cannot access any information on the device—for example, by encrypting the file system. It is then possible to cache some of the network credentials on the device. Ultimately, however, some authentication to the network should always be based on an action or token that is separate from the device.

6.3 Supplementary security

It is possible to augment the security of the air interface by creating a secure path beyond the mobile network either to a specific end point or to the perimeter of a corporate network. In a sense, this always implies a tunnel, also known as a virtual private network (VPN). VPNs were not designed with wireless networks in mind and are therefore more prone to failure due to unreliability and low bandwidth. While the impact of this will be

6.3 Supplementary security

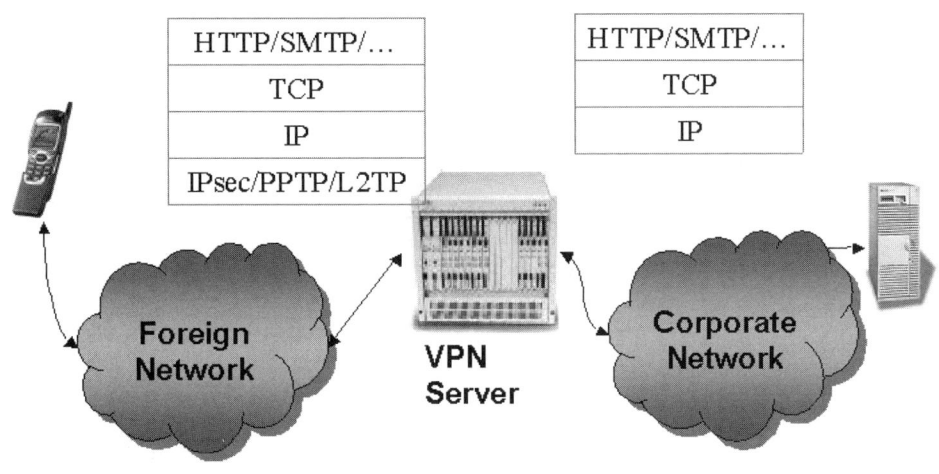

Figure 6.6   *Virtual private network.*

reduced as networks improve, it is a factor that must be considered in any current deployment. (See Figure 6.6.)

There are several different VPN protocols available. Some of the more common ones include PPTP, L2TP, L2F, and IPSec. But obviously both the mobile device and the server must support a common protocol. The limitation is usually found on the device, since it will often not have any VPN client available, and, if it does, it is likely to be restricted in terms of which protocols it supports. It isn't difficult to get tunnel servers for any of these standards.

One unusual VPN, very particular to the mobile environment is WAP, or, more specifically Wireless Transport Layer Security (WTLS). WTLS is based on TLS, the successor of SSL. However, unlike TLS and SSL, which are intended for end-to-end encryption between the client and the application server, WTLS is contained in the WAP stack and only encrypts the path from the client to the WAP gateway. The WAP gateway will often reencrypt the data to the application server using SSL, but the fact that it is decrypted in the WAP gateway implies that there is no end-to-end encryption. WTLS is, therefore, functionally more similar to the VPN protocols than to TLS/SSL. (See Figure 6.7.)

In addition to privacy and integrity, which WTLS always provides, it is possible to stipulate authentication requirements. All WTLS sessions are categorized into one of three classes. Class 1 is anonymous, meaning that neither party is authenticated. Class 2 implies server authentication only.

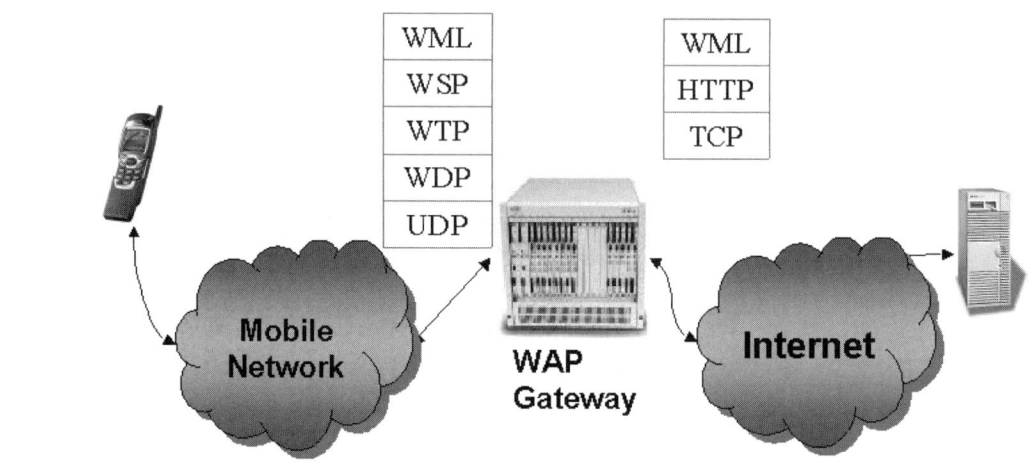

Figure 6.7   *WAP schema.*

Class 3 requires both the client and server to authenticate themselves by providing a signed certificate.

WAP authentication works very much like the network authentication mentioned earlier. Depending on the bearer network we may have PPP over circuit-switched data or we may simply have a packet-data network (such as GPRS). In terms of client authentication this means possibly PPP authentication as well as possibly WTLS (in the case of Class 3) authentication or, maximally, two user authentications. Or, it can mean no authentication at all—for example, in the case of GPRS with WTLS Class 1.

There are, however, some additional considerations. WAP gateways may also be configured to provide an additional level of authentication. This would typically work with the WAP gateway sending the user a WML form that asks for a user name and password. This transaction could optionally be encrypted using WTLS so that the WAP gateway credentials could not be intercepted. (See Figure 6.8.)

That is an additional feature of WAP authentication, and it is very useful since there are also some drawbacks, particularly with the WAP devices currently available.

One big drawback is that since WAP devices are not an open platform it is not possible to influence the authentication procedures or add additional ones. In particular some devices do not support all the common authentication protocols. For example, Nokia phones do not work with MS-CHAP.

6.3 Supplementary security

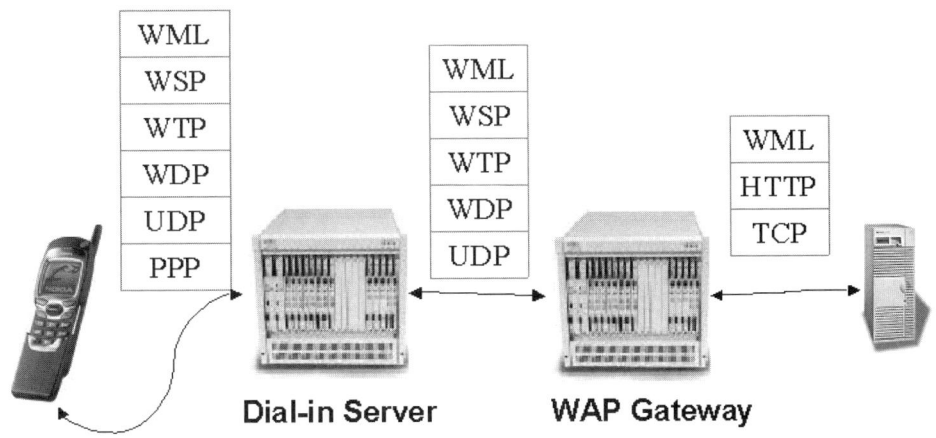

Figure 6.8 *WAP authentication.*

And most of the phones with Phone.com (OpenWave) browsers do not allow backslashes, which makes multi-domain authentication difficult for Microsoft shops.

A further risk to security is the fact that WAP phones often store the credentials on the phone. This opens op the possibility that someone with a stolen phone could extract the credentials and attain unlimited access to the corporate network.

There are two ways to get around this. Either you set up a dedicated WAP dial-in service, which uses trivial credentials and instead relies on the WAP gateway to perform the authentication, or you rely on public PPP servers (i.e., ISPs) to provide the dial-in service and place the WAP gateway in a DMZ, where it acts as a tunnel server to provide access to the corporate network. (See Figure 6.9.)

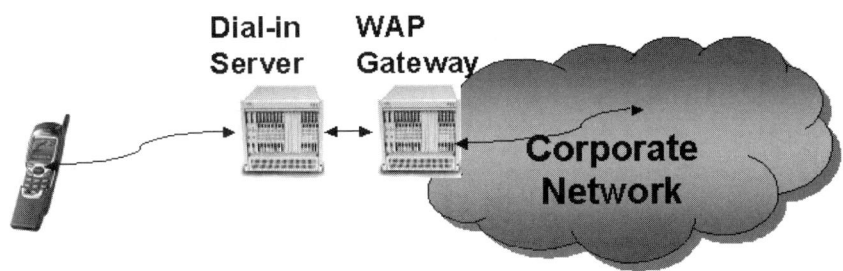

Figure 6.9
Dedicated WAP dial-in service.

In the case of dedicated dial-in service you would need to ensure that the PPP service and the WAP gateway are either collocated on the same machine (with no IP forwarding) or use a dedicated network link. The dial-in service would then only provide access from the WAP device to the WAP gateway. After authentication the WAP gateway would effectively function as a proxy and would relay all the WAP traffic (converted to HTTP) into the corporate network and vice versa.

With this scenario the credentials of the dial-in service would not be sensitive, since they would not be used for any other purpose. The "real" authentication would occur at the gateway, and, even if the PPP credentials were jeopardized, they would not pose a threat to the corporate network.

Another approach would be to allow users to dial in to an ISP and connect from there to the corporate WAP gateway. By placing a WAP gateway in the DMZ it could be protected against many attacks (by only allowing WAP traffic to pass through the external firewall to it). It could also perform authentication of the user. And the internal firewall could be configured to allow only HTTP/SSL traffic from the WAP gateway. (See Figure 6.10.)

Note that both WAP configurations assume that the device is configurable. Many mobile operators (particularly in North America) sell WAP phones with hard-coded configuration settings. The first approach requires that both the phone number of the dial-in server and the IP address of the WAP gateway be configurable. The second scenario requires only a configurable WAP gateway.

If these conditions are not met, then the only recourse is to rely on the security of the mobile operator. I do not recommend any such approach, since it is not possible to verify or enforce any level of security.

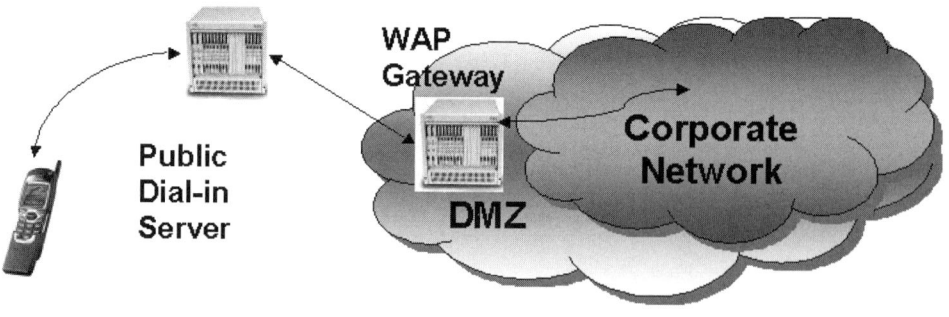

Figure 6.10 *WAP gateway to the DMZ.*

6.4 Enterprise requirements

Enterprises typically have a dual-level security structure. The first level is the perimeter of the corporate network. In order to reduce the threat of industrial espionage or deliberate sabotage, only employees and authorized contractors are allowed any access into the network. While this safety net is difficult to enforce completely, it does thwart the attempts of casual hackers and creates an additional obstacle for sophisticated intruders.

Beyond the common perimeter a second level of security may protect data and applications on an individual basis.

6.4.1 Perimeter security

What does this mean for wireless implementations? First, the secured perimeter must be accessible to mobile devices. Second, access to the perimeter from the mobile device must be encrypted, in order to ensure that it is not intercepted or falsified. Typically, the solution to both of these means the use of a VPN. It is not simple, however, to find mobile devices that support VPNs at this time.

As we try to strengthen client authentication, there are two directions we can take. The first is multi-credential authentication. We can require a password (PIN) to access the device. Then we can require another set of credentials to access the corporate network. And, finally, we can request further authentication from each sensitive application. While cumbersome to the user this approach does permit a tiered authentication scheme, which will reduce the impact of any compromised credentials.

Another dimension of authentication is multi-factor authentication. In addition to merely entering a memorized user name and password ("something you know") we can require other forms of authentication. This can range from various types of removable and/or contactless smart cards ("something you have") to biometric techniques, including fingerprint scanners and voice recognition ("something you are"). A combination of several of these techniques can provide a very effective security scheme.

6.4.2 Application security

The most sensitive applications need to maintain an additional level of security configurations that include authentication, authorization, and auditing. Users who have a business need to access the application must authenticate to the application before they can use it. Depending on their role and responsibilities they may be given different authorization levels (e.g., read-only, modify, delete) or authorization only to certain subsets of data. All the actions requested and performed are logged to preserve an audit trail.

Most of this security is independent of the means of access. If it works for fixed lines, then it should also work for wireless users. The only additional factors that need to be considered in a wireless context are whether the user will require two sets of credentials (for mobile and fixed use), and whether some kind of single sign-on will be supported. Given the cumbersome entry of passwords on some mobile devices, it is not user-friendly to require them to be entered separately for each application. At the same time, mobile credentials are more likely to be compromised, so the risk must also be minimized.

6.5 Secure transactions

Excluding military applications the most sensitive types of transactions are financial. Once money is involved, the incentive is clearly there for criminals to attempt to intercept and/or falsify any transmission. Depending on the value of the transaction the incentive may be quite substantial and can cost justify an expensive attack on a financial system.

The main differences between transactional security and enterprise security are as follows:

- The absence of a perimeter
- The absolute requirement of integral end-to-end security, including privacy, integrity, and, above all, nonrepudiation

A successful implementation of transactional security can serve as a platform to enable any type of wireless financial transactions, including mobile banking, mobile stock trading, mobile commerce, mobile betting/gambling, as well as other nonfinancial but still legally binding transactions such as reservations.

6.5.1 Public-key infrastructure

Most approaches to achieve this security involve public-key cryptography. This does not simply mean installing of a piece of software on the device and the application server. A full public-key infrastructure must be implemented, including certificate directories, certification, and registration

6.5 Secure transactions

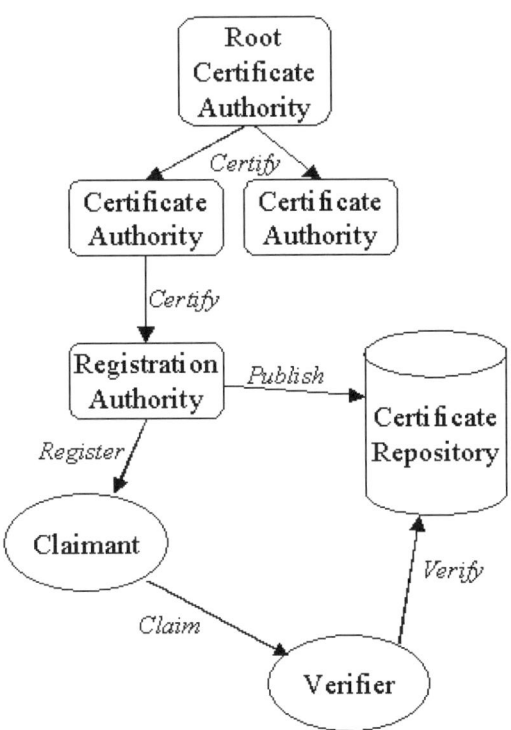

Figure 6.11
Hierarchy of the Certificate Authorities.

authorities, as well as adequate revocation checking protocols. These may be mobile-specific or general-purpose PKIs.

Typically, one or more Certificate Authorities (CAs) are chained together in a hierarchy. This allows only Verifiers to be seeded with a minimal number of Root CAs. They can then verify the validity of the actual CA by validating each certificate in the chain. (See Figure 6.11.)

In addition to the CAs, which mainly produce certificates (although they can also publish them), it is also necessary to register certification candidates. This means verifying their identity, retrieving a certificate from the CA, publishing that certificate in a globally visible repository, and also passing it on to the Claimant.

The Claimant eventually makes use of his or her certificate by passing it to the Verifier. The Verifier can then validate the certificate. This means processing the internal structure of the certificate to ensure consistency, checking that at least one of the CAs in the chain is trusted, and checking the (revocation) status of the certificate—for example, in the Certificate Repository.

This is a simplistic view of a typical PKI. It entails a number of challenges, which must be addressed. The RA must follow careful procedures at registration to ensure that a certificate is not issued to an impersonator. But even if it is correctly issued, there is a risk that the user could abuse it or that it could be compromised.

In some of these unplanned circumstances it is important to define the consequences and procedures that must be followed. Certificate revocation is a complex aspect of a PKI, but one that is very important in order to limit the financial and legal liability of each of the entities.

If we focus on the concept of a wireless PKI, there are several services that can be identified. Server certificates are needed for all the application and infrastructural servers involved in the configuration. Clients will also require certificates if they need to authenticate themselves. The whole certification infrastructure may be independent or joined with a wired PKI, but, one way or the other, the clients need to be configured to trust the root CA.

Further requirements might be the validation of certificates and, in particular, the publishing and accessibility of Certificate Revocation Lists. Payment services are another potential area for wireless PKI services to develop.

Certificate distribution can occur in many forms. Device and WIM manufacturers will install client certificates at time of manufacture, and carriers will issue client certificates to their subscribers. Additionally, content (end service) providers may issue client certificates, particularly if they don't trust the operator to authenticate the user. At the server side, all content providers, as well as potentially mobile operators, and WAP gateway providers will require server certificates.

Once wireless PKIs are available the potential for applications is enormous. Mobile banking, stock trading, and B2C e-commerce are among the most visible, but there are also many other applications. From mobile betting to restaurant, hotel, and airline reservations, there are numerous opportunities.

6.6 Summary

As we have seen, security issues are largely the same whether the environment is mobile and wireless or stationary and tethered. However, there are some additional factors to consider when developing a wireless system. Supplementary security such as virtual private networks and smart cards can help to reduce the vulnerability of mobile devices and even provide a more secure configuration than is commonly found with current fixed devices.

6.6 Summary

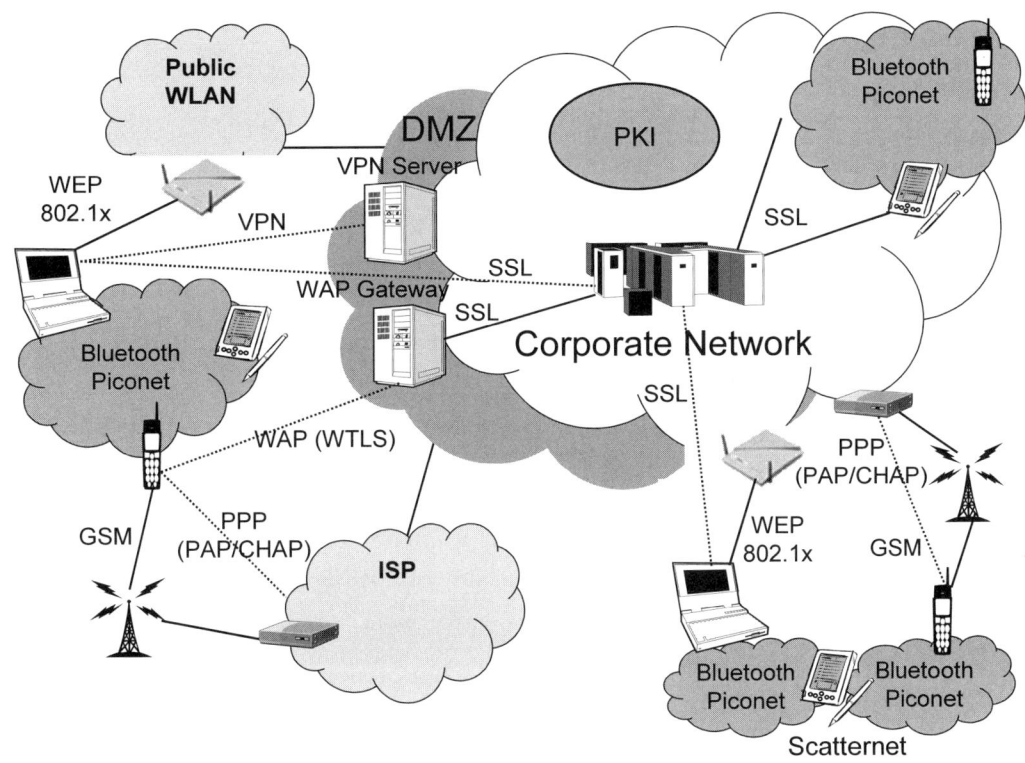

Figure 6.12 *Security mechanisms.*

Figure 6.12 illustrates the interaction of some of the security mechanisms described in this chapter. Clearly there are many possible combinations, depending on the network topologies and applications configurations that must be protected. What is important to realize is that it may be relatively easy to implement a secure pilot of one particular technology. However, when it comes to deploying a full set of solutions on a wide scale, the challenge grows exponentially if we are to ensure that all fronts are secured but that the user is able to operate without undue effort.

Bibliography and related Web sites

General

Burton White Paper on Wireless Security:
http://www.tbg.com/public/doc.asp?docid=244

Bluetooth

http://www.niksula.cs.hut.fi/~jiitv/bluesec.html

http://www.bluetooth.com/developer/download/download.asp?doc=174

WLAN

WECA Analysis of WEP security:
http://www.wi-fi.net/pdf/Wi-FiWEPSecurity.pdf

University of Berkeley study of WEP:
http://www.isaac.cs.berkeley.edu/isaac/wep-faq.html
http://www.isaac.cs.berkeley.edu/isaac/wep-draft.pdf

Smart cards

Rankl, W., and W. Effing. *Handbuch der Chipkarten*. Hanser Verlag, 1999.

PC/SC: http://www.smartcardsys.com/

PKCS#11: http://www.rsa.com/rsalabs/pubs/PKCS/html/pkcs-11.html

Opencard: http://www.opencard.org/

Javacard: http://www.javasoft.com/products/javacard/index.html

7

Implementation

Implementing a mobile solution is similar to deploying any other kind of application. At some stage, after looking at your requirements and deciding whether there is a business need to proceed, you need to architect your solution. This design is the main topic of this chapter.

The architecture involves looking at the building blocks that are available, either through in-house development or for purchase on the market. Once we decide on what components we need to build our solution, the next step is to organize them into a clear and efficient topology. These are the issues we will be looking at.

Obviously, you need to follow a structured approach during the whole project life cycle, including thorough documentation and testing. My omission of some of these tasks should not minimize their importance but rather imply that they are not particular to wireless. This is a book about technology and not project management.

7.1 Architecture/design

Figure 7.1 illustrates the four main components that we need to incorporate in our design, as follows:

- Devices
- Networks
- Gateways, if required
- Applications

We've already covered most of the topics relating to devices, networks, and applications. But we've only briefly touched on gateways so far. Since they fill a critical gap in any wireless implementation, we will deal with them in more detail in this chapter.

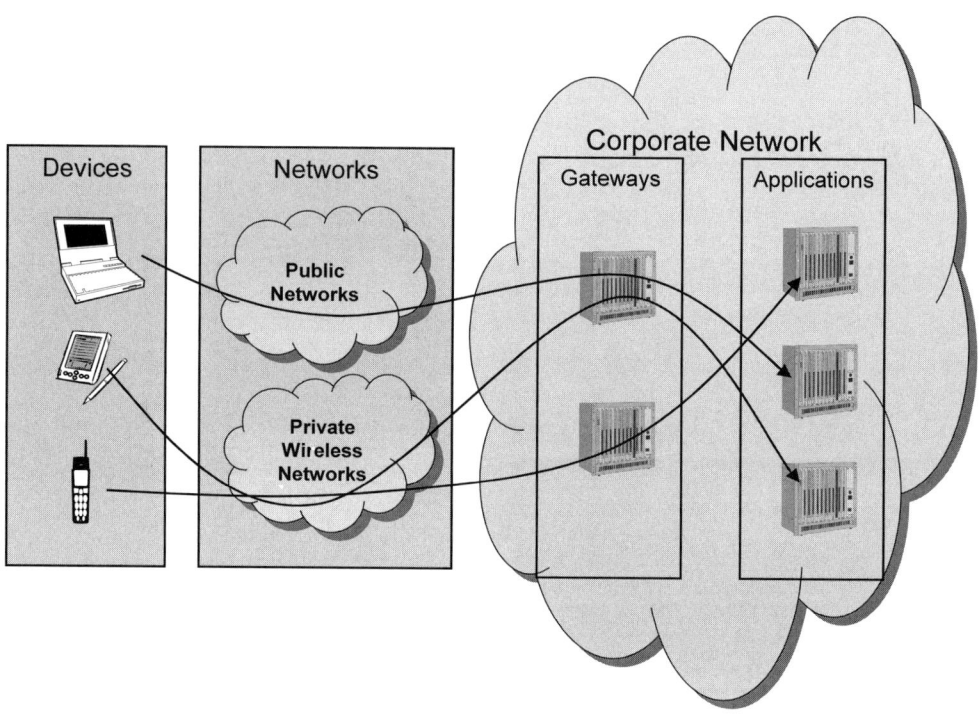

Figure 7.1   *Design components.*

7.1.1   Market landscape

One of the first steps in any implementation is to look at what products and services are available on the market. Most projects will build on these elements to architect a complete solution. The market is very dynamic, but this section will try to give a glimpse of the landscape.

Figure 7.2 illustrates how the content passes through a number of different stages to get to the user. It begins at the content provider and is transformed to meet the format of the user. It then may optionally be collected by a content aggregator and is transferred over the air to the wireless device held by the user.

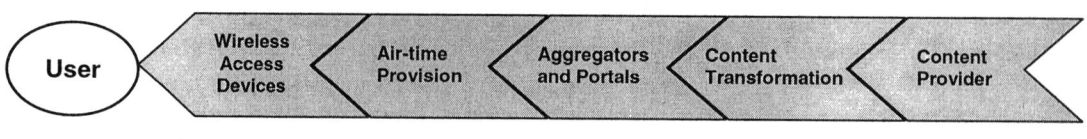

Figure 7.2   *Content stages.*

7.1 Architecture/design   179

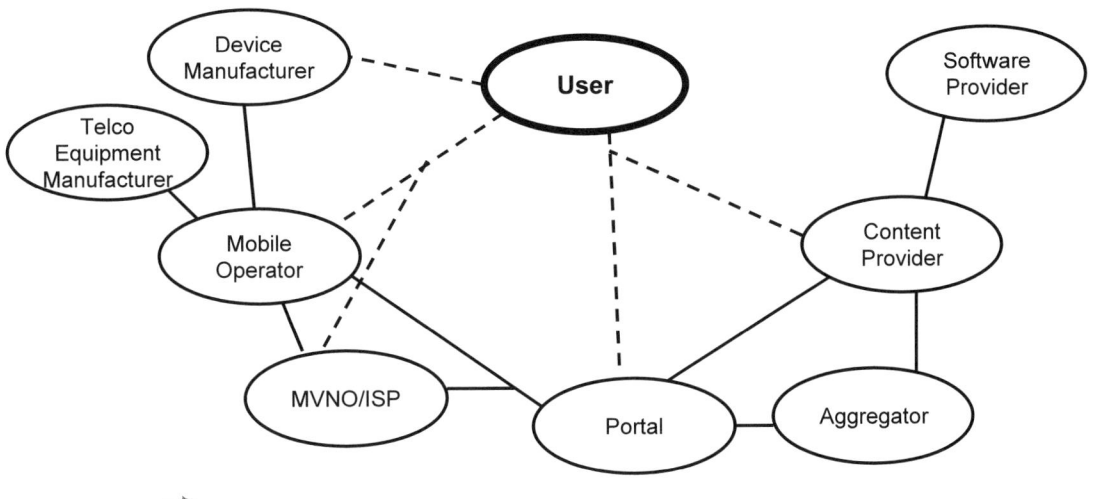

Figure 7.3   *WWAN scenario.*

The path is relatively simple, but behind the scenes even more is happening. Each content type and means of access will involve different players. Figure 7.3 illustrates a typical WWAN scenario.

The market involves many different companies interacting with each other in complex relationships. Fortunately, the user does not need to directly interface with all of them, and in many cases a single salesperson will represent all of them to the user.

7.1.2 Device manufacturers

Device manufacturers deliver both end devices (such as phones and PDAs) and connectivity options (such as PCMCIA cards). We've covered mobile terminals in detail in Chapter 3. There is no reason to repeat that discussion here, but we need to at least briefly mention the main producers.

Phones have the lion's share of the device market. Nokia, Motorola, Ericsson, and Siemens are some of the biggest providers of phones, but there are numerous others, including familiar names such as Philips and Sony. Some of the big names in the market of PDAs include Palm, Compaq, Psion, Handspring, HP, and Casio.

There are more connectivity options than can be listed here. Nokia, Option, and Xircom provide GSM (CSD, HSCSD, and GPRS) PCMCIA cards. Ricochet provides cards for its own interface. Cards are also available for CDMA (SeMax Wireless) and CDPD (Spider).

Table 7.1 *Decision Criteria*

Network Selection Criteria	Examples
Coverage type	In-office, home, hot spots, metropolitan areas
Coverage footprint	Continents, countries, regions, cities
Speed	9.6 Kbps–11 Mbps
Charges	Subscription, connection time, data volume

7.1.3 Networks

On one hand, the choice of networks is a critical decision. After all there are some fundamental differences between PANs, LANs, and the many types of WANs. On the other hand, an ideal network should be completely transparent to the end user. Except in very rare situations, our goal should always be to enable all of our devices to work over any of the networks we choose.

The basic decision is which area we need to cover (see Table 7.1). There are multiple aspects to this. We need to decide whether we need coverage only in the office or whether we will also want network access from the home, a small set of public areas, comprehensive metropolitan coverage, or even coverage in deserted areas.

But beyond the question of which types of areas, we also need to know the geographical extent of the coverage. Do we need global coverage for every country and continent of the world? Can we narrow this down?

WWAN versus public WLAN

The mobile phone operators and paging networks were the first to offer wide area coverage. And for that reason they are often called Wireless Wide Area Networks (WWANs). However, the term is misleading, since the range of each base station is limited to a few kilometers.

As we have see in previous chapters, in order to provide extended coverage they require a network of base stations. But this technique is not reserved to mobile phones or pagers. In the past two years, there have been a number of efforts to provide networked WLAN coverage.

While the approach is similar on the surface, there are typically a number of practical differences in the two solutions, including the following:

- Data throughput—Wireless WANs can currently accommodate less than 100 Kbps of data throughput. Contrast this with up to 11 Mbps available on WLANs.

- Cost—Mobile phones are also typically more expensive to use than WLANs. The former often bill based on air time or data volume, while the latter frequently offer low-cost subscriptions providing unlimited use.

- Coverage—Given the longer range of WWANs they are more likely to provide coverage in remote and sparsely populated areas. WLANs are often confined to "hot spots" such as hotels, airports, and offices. But there are an increasing number of attempts to provide coverage for entire metropolitan areas.

- Roaming—WWANs (particularly GSM) can provide international coverage through roaming agreements. These are currently uncommon for WLANs. However, WLAN operators do not require a license, so they may freely operate in many different countries. It is also possible to seamlessly use the same device with multiple WLAN configurations. This is not yet possible with WWANs.

The list of providers is continually growing and includes many well-known names, such as 3Com, Cisco, Compaq, Dell, Fujitsu, IBM, Intel, Lucent, Nokia, Proxim, Samsung, Siemens, Sony, and Toshiba.

Some of the primary considerations for WLAN deployment include pricing, range, and reliability. But it is also important to look at the ease of installation and management, security features, and scalability. In order to protect your investment you will also want to look at multi-vendor interoperability and the product road map.

Mobile operators

Mobile operators represent an out-sourcing of wireless infrastructure. They typically service millions of subscribers. The marketplace for mobile carriers is very dynamic with mergers and acquisitions an everyday occurrence.

In some cases the customer does not directly interact with the mobile operator. Instead, multi-vendor network operators (MVNOs), otherwise knows as Wireless ISPs, buy air time from carriers and then resell it to their customers. Some examples of these MVNOs include Palm.net, GoAmerica, and OmniSky.

The operators do not typically develop their equipment themselves. Instead they rely on Telecommunications Equipment Manufacturers (TEMs), such as Nortel, Ericsson, Nokia and Motorola to provide the infrastructure they need. These TEMs also draw on other manufacturers, such as Hewlett Packard and Compaq, to supply parts as well as components for managing and billing.

Public operators are not exclusive to Wireless WANs. Using 802.11b technology MobileStar and SkyNetGlobal have begun to build an international network of subscription-based WLANs. Some of the cell-phone operators, such as Finnish Sonera and Swedish Telia, have also begun to deploy WLAN services to complement their WWAN technologies.

When selecting a preferred mobile operator for your users, there are several factors to consider, including the following:

- Coverage
- Air interfaces supported
- Transmission quality
- Pricing
- Support
- Devices
- Supplementary services

Coverage

Coverage has multiple dimensions to it. It includes the regional or national footprint of the operator's presence, such as the country of Austria or the Northeastern United States. But within that area there may not be comprehensive coverage. There is usually a concentration of base stations in the metropolitan areas, but motorways and small towns may not all be equally well served. As the population density decreases, so usually does the coverage. Mountains, deserts, and lakes are also difficult or expensive to cover and provide relatively little benefit.

The coverage area can be extended through roaming agreements with other operators. GSM offers some of the most extensive opportunities covering the majority of the world. But here it is important to note that not all GSM carriers have mutual roaming agreements. And since there are three frequency bands available for GSM, it is important to have a terminal that is compatible with the host operator.

Air Interfaces

The set of available air interfaces is an important factor for determining which applications will be functional. It is critical to understand not only which data interfaces are supported (e.g., HSCSD, GPRS), but also which standards will be available in the future and what the time frame for this evolution is.

It is also worth verifying whether any roaming agreements include these interfaces, particularly if the users are likely to travel frequently.

Transmission quality

The transmission quality typically varies with the number of active users in a given area and is directly impacted by the density of the base station available. As such it is difficult to compare two operators on a broad scale. Nonetheless, it is important that the preferred carrier be able to provide good quality at least in the home offices and in the general vicinity of any other typical business operations.

While this book is primarily about data transmissions, we need to remember that voice is the primary wireless application and will be for some time to come. This means that we need to ensure not only good data rates for each air interface but also clear voice transmissions.

Pricing

The pricing structure is often difficult to compare without a good idea of use patterns. There are at least three components to the total charges: subscription, air time, and data transmission volume. These must all be added together based on estimated average use.

It is often possible to reach an operator agreement, not only for a discount based on heavy use but also to provide special services. A dedicated base station on the customer premises can be restricted to employees and result in no, or at least reduced, charges for local voice and data transmissions. It also has the advantage of lower latency and faster circuit setup times.

Support

Operator support is essential, since phone and data access are business-critical operations. This covers everything from the average hot-line wait to requesting preferred corporate hot-line treatment, of utmost importance for special operator agreements.

Devices

It is often possible to purchase phones and other devices from/through the operator at discounted prices. However, it is important to verify what this entails. These terminals are typically locked to the operator (if there is no SIM card, they are always locked), which implies a long-term commitment to the agreement.

Supplementary services

Many operators also offer supplementary services. These can encompass almost any type of application, including WAP and location-based services.

7.1.4 Topology

In addition to selecting the components, we need to decide how to put them all together. This means quantifying and placing the infrastructural components (access points, application gateways, WAP servers, dial-in servers, VPN servers) in order to maximize availability and minimize cost.

We need to consider both geographical coverage and, in many cases, international interoperability issues. We also need to choose the types of interfaces we require, whether they by publicly or privately operated. And, finally, we need to ensure that all of our mobility needs are met. Does our infrastructure support users regardless of their location in our network? Can the applications cope with dynamic network address in the course of a session? Can the infrastructure handle users in motion.

WWANs

Wireless Wide Area Networks don't require much wireless-specific infrastructure. Most of the equipment is supplied by third parties such, as mobile operators. What is important is the placement of VPN and dial-up servers. Although not dedicated to wireless traffic, their use will increase and become more critical as mobile access grows.

This means that it might be necessary to increase the number of access servers and ensure that they are geographically distributed to best serve the locations of the users. More servers (with good failover and load balancing) will increase the reliability and overall throughput of the system.

Proximity of the servers to the users will result in lower latency and therefore better response time. In the case of dial-up servers, it can also reduce the users' phone charges if there are servers available in the local area.

Corporate WLANs

A corporate Wireless LAN (WLAN) means privately owned and operated infrastructure and obviously entails very careful planning for a solid implementation.

Standard 802.11 supports both a peer-to-peer configuration called ad hoc mode and also a network configuration called infrastructure mode.

Ad hoc mode

An ad hoc network is simplest to set up and is therefore typically used in homes or small offices. It contains no access points. Instead, all stations communicate with each other directly in a symmetric relationship. The coverage area is limited by the range of the individual stations. (See Figure 7.4.)

Infrastructure mode

A network operating in infrastructure mode requires an access point (AP). The AP regulates all data transfer between the individual stations and a so-called distribution system, such as a wired LAN.

The relationship between the AP and the stations is asymmetric. All data, even between stations, must pass through the AP.

Infrastructure mode is the preferred configuration for enterprises and large offices, since it offers more direct control and manageability. It is able to accommodate configurations with both wired and wireless LANs and also supports more extensive security and encryption settings. While it does require at least one access point, the cost of the AP becomes less significant when there are a large number of stations. (See Figure 7.5.)

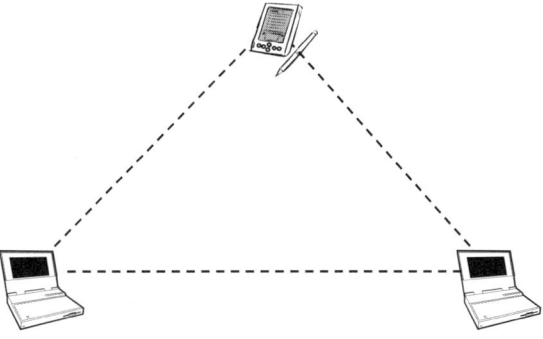

Figure 7.4
Ad hoc mode.

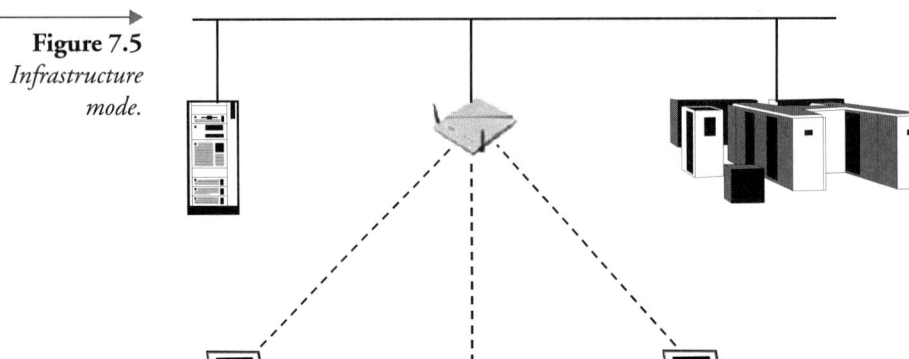

Figure 7.5 *Infrastructure mode.*

Since wireless LANs must often span areas that exceed the range of a single base station, 802.11 incorporates the ability to employ multiple access points. In this case each access point and its associated wireless devices is called a Basic Service Set (BSS).

An Extended Service Set (ESS) is a collection of BSSs in the same network that work together to support roaming of devices from one AP to another. (See Figure 7.6.)

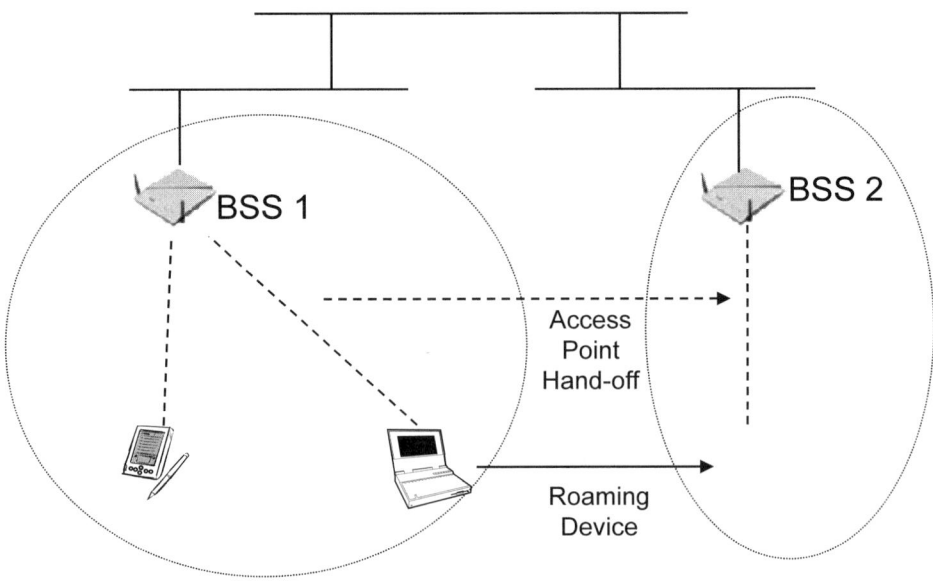

Figure 7.6 *Extended service set.*

While implementing 802.11 in a small office may be a very straightforward task, it can be considerably more difficult in a large facility. It may be necessary to conduct a careful site survey to ensure that the APs are optimally placed in order to maximize coverage and quality of reception. If the facility uses more than one IP, subnet planning is even more crucial as the ability to roam is significantly restricted, which may impact many of the applications.

Determining the optimal placement of access points Depending on the size of the area, placement of access points can be trivial or extremely difficult. A home or small office can often operate with one centrally positioned access point, requiring very little planning. On the other hand, a large office building or manufacturing plant will require many different access points and positioning them is certainly a challenge.

Unfortunately, there are very few tools available to help with the task. Whether you roam the area with your laptop, employ a signal strength indicator, or try to simulate the environment with mathematical models, you can never achieve a high degree of precision.

There are simply too many variables, ranging from the construction of the walls, and placement of elevator shafts to other sources of interference, such as microwave ovens. Our own mobility also provides an unpredictable variable to model. Human beings consist mostly of water, which can absorb a substantial amount of energy. And, of course, a few dozen 802.11b-equipped laptops in a conference room may run into problems if they all try to view streaming video at the same time.

While a small office or home may be able to get away with an ad hoc topology, it cannot scale to the size needed by a major corporation for ongoing operations. An infrastructure-mode topology is a more complicated task, not only because of the requirement for placing access points but also because of its intricate relationship with the wired-network topology.

If access points in an ESS are placed on different subnets, many applications will have difficulty roaming from one AP to another. It is, therefore, very important to develop an integrated network design.

Public WLANs

Public WLANs are very similar to WWANs in terms of implementation. They also require VPN servers to access a corporate network. But Public WLAN providers also differ in the types of services they provide and the way they charge for their access.

Public WLANs in confined buildings (e.g., hotels, airport lounges, convention centers) may be able to provide access to printers and fax machines or, increasingly, voice calls using VoIP (Voice over IP) and streaming video.

These services, as well as simple access to the Internet, come at a cost. Whether these are corporate or individual subscription, they can be charged at a flat rate, based on connection time, the amount of data transmitted, or the services used.

7.1.5 Aggregators and portals

Portals provide functionality similar to the interface you find on many Web search engines, such as Yahoo!. They include links to many sites, providing related information that is often personalized by the user to ensure relevance.

Aggregators go one step further by embedding the content from several external sources into one unified stream. They send users highly personalized Internet information that they can view on any device, regardless of whether their carrier has an alliance with those content sites.

Some of the market leaders in this area include OracleMobile.com (a subsidiary of Oracle) and Strategy.com, (a subsidiary of MicroStrategy, a data-warehousing firm). They allow users to select information they want to receive on their wireless devices from corporate Web sites.

Yodlee allows consumers to aggregate information from Web sites where they have an account, such as their bank or their brokerage. Originally launched as a Web-based service, it now allows consumers access to the data via a WAP-enabled cell phone. Plumtree, an established Knowledge Management portal, has also begun to offer wireless services to its users. Even Microsoft is active in this space with its MSN Mobile offering.

7.1.6 Middleware

Wireless application gateways, also known as middleware, are a new topic. They are difficult to cover, since they represent different kinds of functionality depending on the vendor. Their main purpose is to facilitate access from mobile devices to legacy applications.

In a wireless environment there are multiple network protocols, device types, and legacy applications. These need to be "glued" together in some fashion. If we take a closer look at some of these components, we can see the problems more clearly.

There are several types of network protocols, including circuit-switched data protocols (such as GSM CSD, HSCSD, IS-95, and IS-136) and packet data networks (such as CDPD, PDC-P, GPRS, and, eventually, EDGE and UMTS). The challenge is not only the diversity of network protocols, but also the fact that many of these are unreliable and none of them provides failsafe security or optimal compression.

Devices come in many different shapes and sizes, ranging from cell phones and pagers to handhelds and laptops. Even within these categories there is an enormous variety in the machine interfaces.

Legacy applications provide completely different interfaces that are not yet tailored to the mobile environment. They often presuppose a powerful desktop with a high-bandwidth connection.

In order to pull together these disparate components we need to add functionality, such as end-to-end security, synchronization of data for off-line use, notification of important events, and content transformation to suit the form factor and machine interface of the device.

We can also facilitate the user experience by providing the capability of single sign-on and storing users' preferences and personal information.

These functions can be broadly classified into three different categories, as follows:

1. Transformation

2. Session control

3. Profile management

Figure 7.7 illustrates the challenge of transforming data from a wide variety of sources, ranging from enterprise frameworks, such as Java 2 Enterprise Edition (J2EE), CORBA, SOAP, and Microsoft .NET, to databases (SQL, SAP) and legacy systems (CICS) or the more future-oriented XML standard.

These need to be transformed into a format that is suitable for a mobile device, including the many mobile markup languages (e.g., WML, HDML, cHTML), SMS messages, or even nonmobile formats, such as HTML and SMTP.

But the transformation is not purely a syntactic task. It should also involve reformatting to suit the target device characteristics, filtering of unnecessary content, and even the inclusion of defaults based on personal preferences.

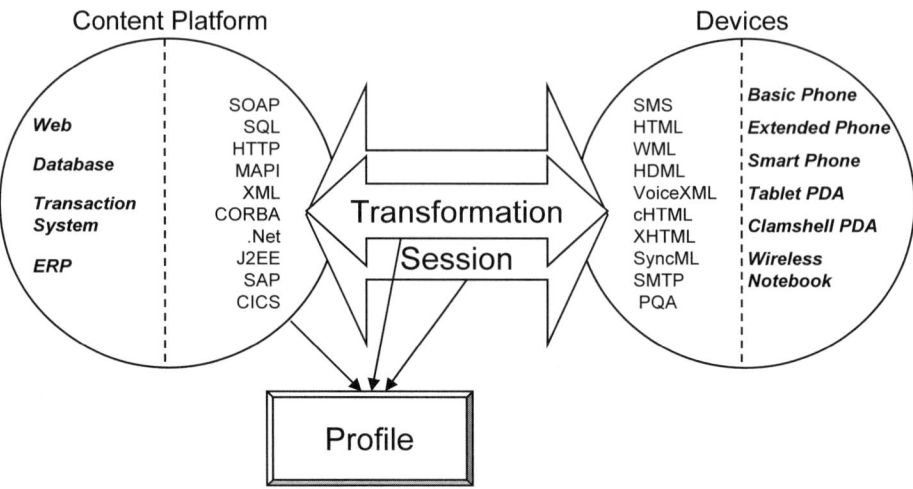

Figure 7.7 *Transforming data from various sources.*

Session control can involve the capability for single sign-on, reestablishment of interrupted communications, session-to-session context preservation (e.g., cookies), identification of the device type, end-to-end security, and synchronization of application data.

A profile is necessary to simplify the authentication process and to provide storage of preference and personal information. It can also provide a mechanism for use-based billing.

Wireless application gateway providers

It is difficult to contrast the many middleware providers because of the varied objectives each one pursues. Any comparison inevitably runs into the problem of comparing apples with oranges.

At a high level, we can categorize providers by their target market. Some are trying to sell to enterprises for Business-to-Employee (B2E) applications. Others are targeting carrier, financial institutions, and Business-to-Consumer (B2C) services that need to provide access to a much larger and diverse public.

In the corporate scenario illustrated in Figure 7.8, the enterprise typically hosts its own application gateway and allows its employees to access corporate (and often also third-party) data. These data may or may not be heavily protected, but the application gateway is only accessible to users who have authenticated to the corporate network.

7.1 Architecture/design 191

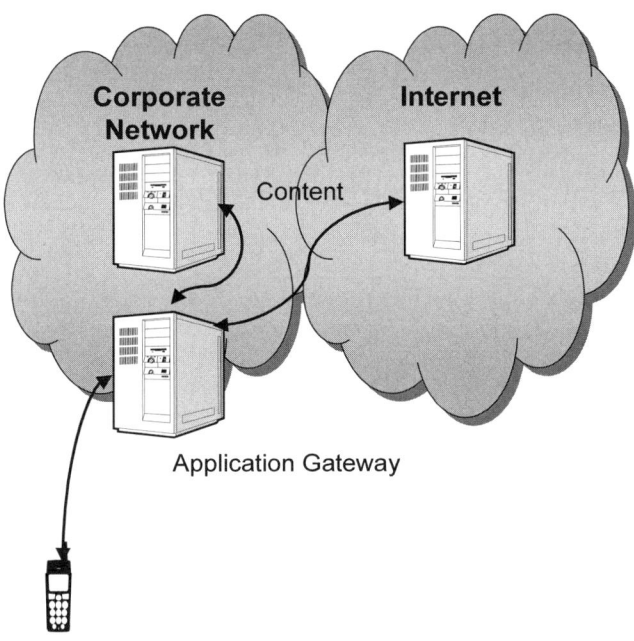

Figure 7.8
Corporate scenario.

When a carrier hosts an application gateway, it can provide user access to its own data and third-party applications (which might include contractual arrangements with enterprise customers to provide employee access). This presents a much more complex scenario, both in terms of the diversity of users and the required functionality in areas such as billing and security. (See Figure 7.9.)

Another major difference is in the back-end services they are providing. These are open-ended and include e-mail and databases. But the more challenging tasks typically involve a tight integration with the individual business processes of each company. This implies connectivity to the .Net Framework, Java 2 Enterprise Edition, CORBA, and even some of the legacy mainframe systems such as CICS.

Middleware providers include big names such as Microsoft (Mobile Information Server), Oracle (Oracle iAS Wireless Edition), and IBM (WebSphere), but there are also a large number of smaller companies, often start-ups and niche players. Some of those receiving recent attention include Aether, AvantGo, Brience, Everypath, Extended Systems, Netmorf, Seagul, and Viafone.

There is little commonality among these products, making it very difficult to compare them fairly. In order to evaluate them for a particular need,

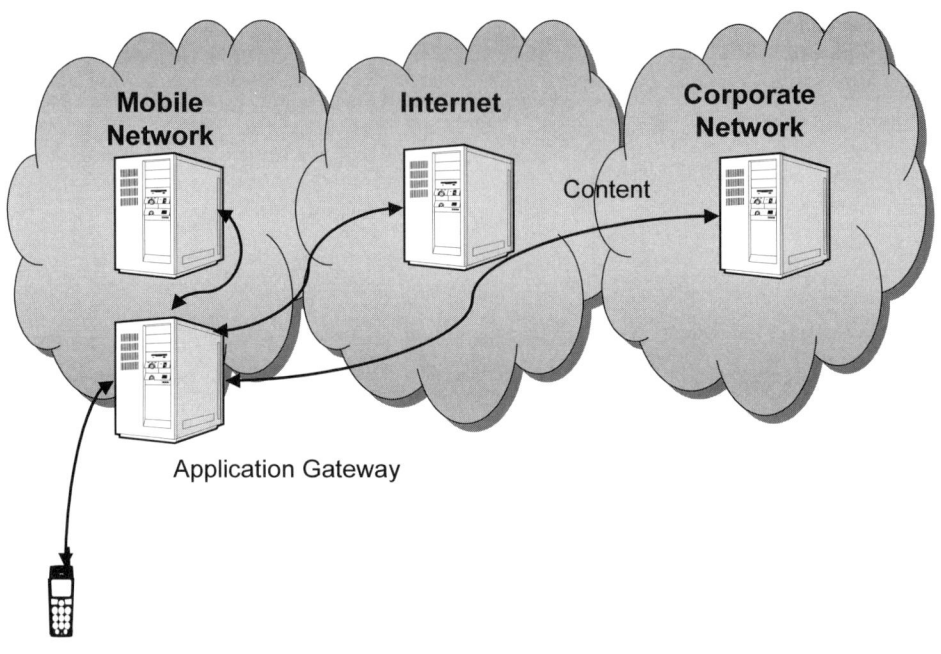

Figure 7.9 *Corporate scenario with a mobile network.*

it becomes important to develop a critical-feature checklist, which might include items such as synchronization, single sign-on, voice integration, transactional support, and XML compatibility.

7.1.7 WAP

As we saw in Chapter 4, there are three primary WAP components to most WAP implementations. They are as follows:

1. The (PPP) access server
2. The WAP gateway
3. The content servers

In a corporate environment there are also three different networks that can host these servers, as follows:

1. Mobile network
2. Corporate network
3. Any third party on the Internet

Figure 7.10
No required mapping of servers to networks.

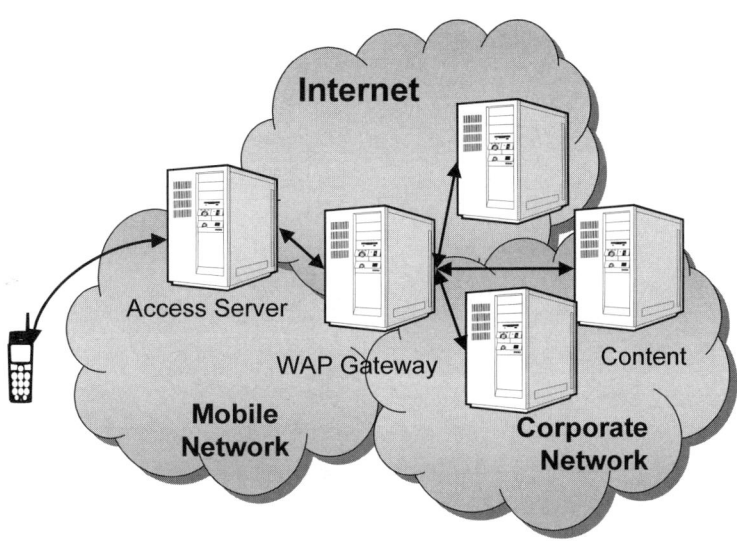

The ambiguous boundaries in Figure 7.10 depict that there is no required mapping of the servers to the networks. Who hosts which server is, at least in theory, at your discretion.

As depicted in Figure 7.11, the WAP gateway (WAP proxy) can be hosted on any of the three networks. If the gateway is hard-coded on the devices, then there may be no other choice but to use the gateway of the operator. However, this implies a certain amount of trust toward the carrier, which may not be acceptable. Devices can always access content on the network of the gateway and typically also on the public Internet. However, they are not likely to be able to access content on the mobile network or a corporate network unless the WAP gateway is hosted there.

In cases where data are circuit switched, a dial-up server must also be available. It, too, can be operated by the company, the operator, or any third party on the Internet (any ISP). Typically the WAP gateway and any access servers are placed on the same network. But, as illustrated in Figure 7.12, it is also feasible to use a public WAP gateway from the mobile or corporate network. Other combinations are very uncommon.

As illustrated in Figure 7.13, there are numerous options regarding who can host which components, each with its own implications in terms of trust and cost.

There is also no recommendation as to whether the customer should be able to select a particular gateway/content provider or if these are hard-

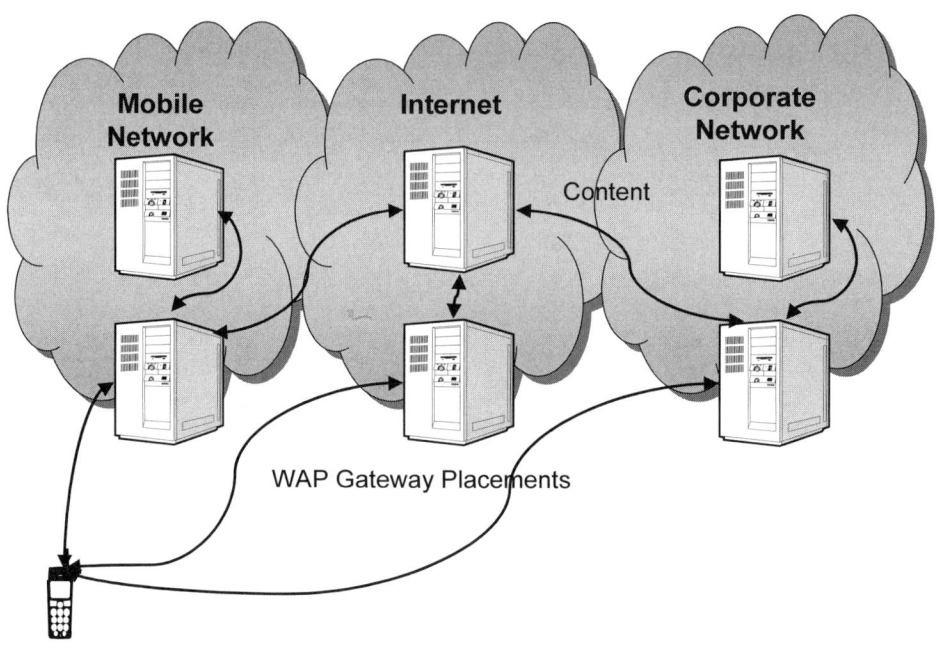

Figure 7.11 *WAP gateway hosted on the three networks.*

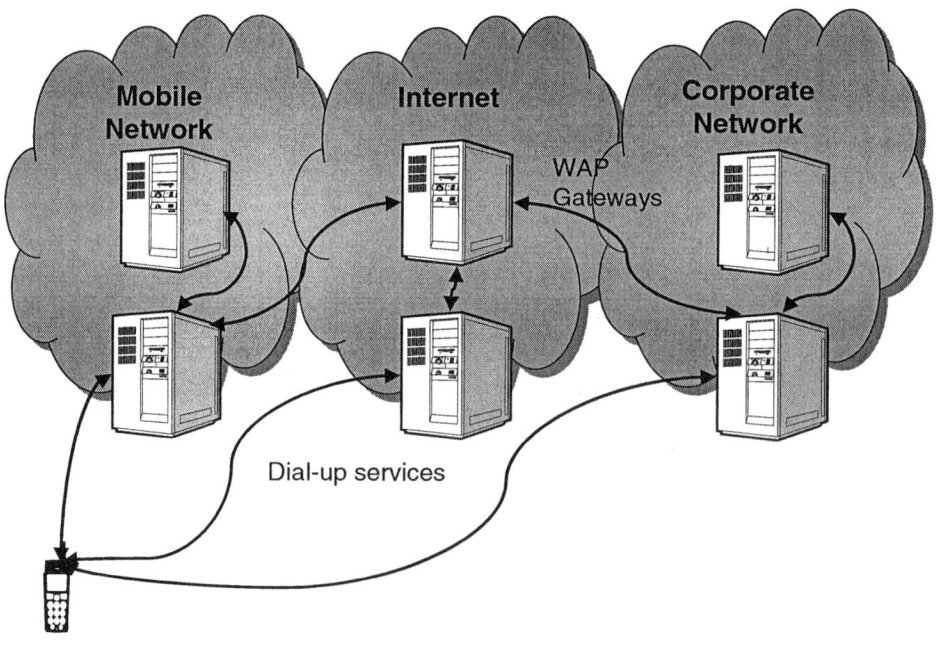

Figure 7.12 *WAP gateway with dial-up services.*

7.1 Architecture/design

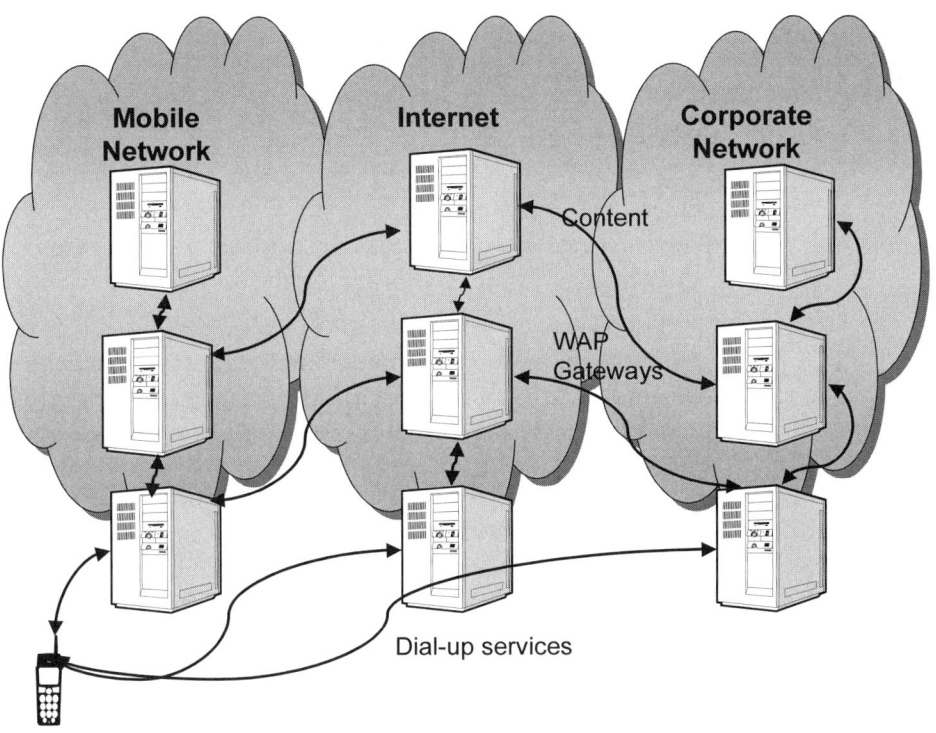

Figure 7.13 *Other host options.*

coded by the network operator. While the consensus seems to be that ultimately an open environment (such as the Internet) would be better, this may not necessarily be the case, particularly in the short term.

WAP gateways

There are quite a number of WAP gateways on the market. Openwave (Phone.com, Unwired Planet) has strong recognition. Some of the others include CMG, Dr. Materna, Nokia, 724, and WAPlite. There is even an open-source version available on Linux, called Kannel.

Some of the factors to consider in choosing the gateway include the following:

- Cost
- Billing support
- Hardware platforms
- Scalability

- Security
- Application integration support

7.1.8 Bluetooth

Bluetooth is a technology not typically deployed by enterprises. The corporate challenge is to ensure that the existing infrastructure (802.11b) will be able to cope with the increasing number of Bluetooth devices.

However, it is also important to realize that the full panorama of future Bluetooth applications is still not known. There are some interesting potential applications of Bluetooth as an access point in competition with 802.11 that might be viable. It is also conceivable to use Bluetooth for very special purposes, such as location beacons, security devices, and so on.

7.1.9 Content providers

The list of providers of wireless content is virtually limitless. In time it is likely that almost all applications and content sources will cater to mobile users. Aether, Everypath, and AnyDevice have already begun to develop a wide range of remote business applications. Some very successful applications, such as Siebel and Oracle, are well on their way to provide interfaces for all mobile platforms. The trend is for all traditional providers, be it news, weather, sports, stock quotes, or even e-commerce, to extend their reach to wireless environments.

7.1.10 Wireless Application Service Providers

In many cases a company may not want to install and manage its own wireless gateways. It is a complex task and many corporations may not have the necessary skills in house, or they may wish to dedicate the resources to other projects.

As an alternative they can outsource their wireless operations to a specialized company, known as a Wireless Application Service Provider (WASP). There are two approaches to this: Either the provider can install and manage the gateway on the customer's own premises, or it can host the gateway and only provide the services to the customer.

What does an Application Service Provider do?

An ASP manages outsourced customer applications. These are typically hosted by the ASP, which takes on all related responsibilities such as deployment and management of the software. It usually operates according to a service-level agreement that relieves the customer of the associated workload and need for specialized expertise while at the same time guaranteeing high service reliability.

ASPs are particularly suited to small and medium enterprises (SMEs) due to their economy of scale. It would not be cost efficient for small companies to invest dedicated resources to installing and maintaining many small applications. But an ASP hosting many thousands, if not millions, of users can afford specialists dedicated to ensuring high availability of all provided applications.

A WASP will usually include wireless infrastructure (such as WAP gateways or WAGs), but it can also constitute a general-purpose ASP and provide access to end applications, such as e-mail.

A WASP may be the only wireless option an SME can afford, but it also represents an easy way to get started for large enterprises—before they invest heavily in their own infrastructure, they have a way to get their feet wet and can experiment with different applications to determine how useful and valuable mobility is to their users. WASPs already have servers, connectivity, management tools, application development skills, carrier contracts, and IT people in place. Setting all that up would take some time and a heavy investment, even for a large enterprise.

A wireless ASP can provide different types of functionality. At the personal level, companies such as Oracle and AvantGo provide stock quotes, sports scores, productivity tools, games, and small-office applications. Providers such as MobileSys and Senada extend these to collaborative applications, such as messaging, calendaring, and e-mail. Aether and AvantGo take on the full range of enterprise applications, including customer and employee relationship management, sales force automation, and e-commerce.

7.1.11 Internationalization

Most smaller companies do not need to worry about international implications. Even large corporations may choose to pursue independent solutions in different geographical areas. But when users travel frequently between countries, it is worth considering a unified approach.

This means ensuring global access to VPN and dial-up servers for all users where it is possible and cost effective. It may also involve ensuring that the devices, and their air interfaces, are able to function in different countries.

Table 7.2 *Channels Available in Various Countries*

Geographical Area	Standards organization	Permissible Channels	Suggested channels
Europe (except France and Spain)	ETSI	1–13	1, 7, 13
France		10–13	11
Spain		10 and 11	11
United States	FCC	1–11	1, 6, 11
Canada	IC		
Japan	MKK	14	14

Given the diversity of phone standards, there is no device that works in Japan, Europe, and North America. However, there are triband phones and PC cards that do provide GSM access to a very high percentage of the world. As more multi-mode phones come on the market, the challenge of matching the air interfaces to their geographical footprints becomes more important.

Even in the case of wireless LANs, which operate in the 2.4 GHz band throughout the world, there is not always seamless roaming, since different countries use different channels.

As shown in Table 7.2, it is not possible to use any single channel everywhere. However, you can use a common channel for your intended coverage and ensure that you have a mechanism for your users to automatically (or at least easily) switch channels when they travel.

7.2 Development

There are two ways to develop mobile applications. Either you can provide content on the server and deliver it to the device—for example, through a generic protocol such as WAP and HTTP—or you can create an application that is specific to the mobile device and uses an application-specific protocol (e.g., SMTP, IMAP4, SQL) to communicate to the server.

The primary markup languages relevant for mobile content are HTML and WML. While WML is clearly a new format, which must be developed for mobile devices, HTML interfaces are already available. This does not mean that all HTML interfaces are tailored to small form factor screens. Some tools, such as the Microsoft Mobile Internet Toolkit, will assist with making your applications automatically render to the user's device type in real time.

There is no rule as to when to adopt which approach, and it depends very much on the application, the devices available, and the needs of the user. It may even be necessary to develop some applications using a combination of both.

7.2.1 WAP

In order to develop wireless content a programmer will require a development environment as well as the ability to test the applications. There are many similarities between conventional Internet development and wireless development. Both involve creating Web (HTML/WML) pages and placing them on a Web server (e.g., Netscape, Apache, Microsoft Internet Information Server).

Testing wireless content will ultimately require the establishment of a complete wireless configuration, including a WAP gateway, access server, and the full set of prospective devices. However, it may not be expedient to perform a complete test at every step during the development cycle. In order to facilitate the testing there are also WAP browsers on the market that are able to simulate wireless devices and connect to the content servers over wired connections, either directly or via a WAP gateway.

Some of the development kits currently on the market include Ericsson, Openwave, and Nokia, which can be downloaded from the Internet. They each provide simulators for their respective devices. Figure 7.14 shows part of the Nokia simulator. The others are very similar in functionality.

7.2.2 Microsoft Mobile Internet Toolkit

The Microsoft Mobile Internet Toolkit is essentially a set of ASP.NET mobile controls, which are available at Visual Studio.NET. It is built on the .NET framework and uses the ASP.NET Web development model.

The controls provide an Adaptive User Interface (AUI), which renders support to a set of mobile devices. Developers program the controls using device-independent properties, methods, and events. When a supported device requests a Mobile Web Form page, the page and controls automatically produce a rendering suitable for the device.

 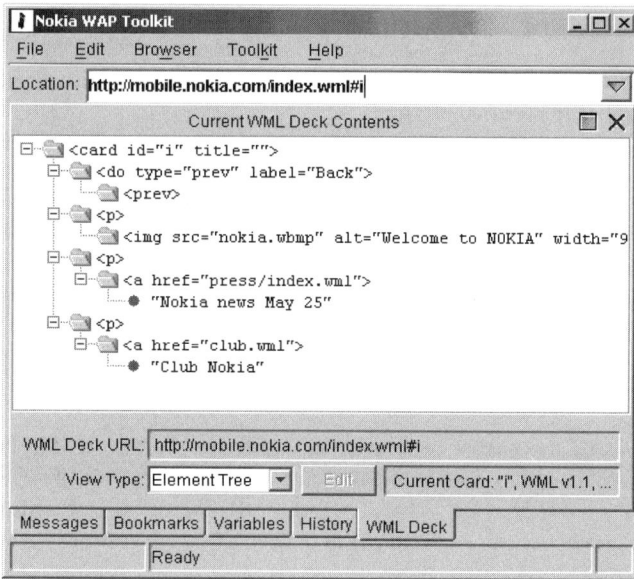

Figure 7.14 *Nokia simulator.*

The object model of a mobile control is device independent so that interaction with it is uniform, regardless of the target device. Web Form programming uses VB programming mode to rapidly construct desktop applications, which means that no knowledge is required of WML or any other markup language. The controls handle all the implementation differences between browsers, gateways, and networks. In addition, it also supports unique mobility scenarios, such as the ability to push content to supported devices.

7.2.3 Platform Development

Windows CE

Windows CE offers two tools for developers. Platform Builder offers solutions for packaging Window E-based embedded systems. Embedded Visual Tools provide a development environment using Visual BASIC and Visual C++.

The Platform Builder supports several different hardware architectures. The microprocessor list is extensible, but out of the box it has support for ARM, StrongARM, x86, PowerPC, and MIPS. It offers the ability to choose from eight configurations of Windows CE, ranging from a simple kernel to a full system with a graphical user interface.

The Platform Builder also includes a number of compilers and connectivity options, as well as comprehensive debugging support including the following:

- A status monitor to view information about communication between the development workstation and target device.

- Remote performance monitor to observe the target processor, network, thread, and process performance.

- Remote system information to see system memory, power, and peripheral and device driver details.

The Embedded Visual Tools provide a graphical development environment based on Visual Studio. It is especially tailored to programmers already familiar with Visual BASIC and Visual C++.

Many programs developed for the desktop can be ported to the mobile devices. However, this is not usually a good approach, since desktop applications are typically less efficient and tuned to a different machine interface. Nonetheless, the uniformity of the development tools makes it easy to learn the Embedded Visual Tools.

Most functionality is available, including the basic COM infrastructure, ActiveX Controls, and ActiveX Data Objects. There are also some additional controls (CommandBar and MenuBar) and support for voice and handwriting recognition and the ActiveSync architecture.

Palm OS

A Palm application consists of two parts. The handheld portion resides on the device and interacts with the user. The conduit portion is an executable library that runs on the server during synchronization (called HotSync in Palm terminology).

To assist with application development, Palm offers a Palm OS Emulator (often abbreviated as POSE), shown in Figure 7.15, that can be downloaded for free from the Internet. This is not quite as simple as it sounds, since the emulator cannot run without a ROM image and in order to get that you need to register for a license with the Palm Resource Pavilion in writing—a process that can take several weeks.

If POSE does not suit you, there are other integrated development environments available, such as Metroworks CodeWarrior, Pendragon Forms, and Satellite Forms.

Palm applications must be written to work with very limited memory. They are event driven and usually developed in C, C++, or Java. Forms have

Figure 7.15   *Palm OS Emulator.*

all the typical UI items, including menus, lists, buttons, checkboxes, and fields. There is even support for a primitive database.

Each Palm application can also make use of one Conduit DLL. It takes care of the server-side synchronization and is responsible for converting the Palm application's database records to the equivalent data structures on the server.

EPOC

EPOC applications are written in C++ or Java. It is possible to use Microsoft Visual C++ as the IDE. The application can then be tested using the emulator in the downloadable EPOC C++ SDK.

The graphical user interface defines dialogs (similar to forms), buttons, lists, editors, toolbars, menus, and shortcut keys, all very similar to Windows and the other mobile platforms. (See Figure 7.16.)

The communications support of EPOC includes RS232, dial-up TCP/IP, IrDA, and a CF slot. Symbian also provides a generic product (EPOC Connect) that provides a complete range of data synchronization and connectivity features, including file management, backup and restore, remote

7.3 Management

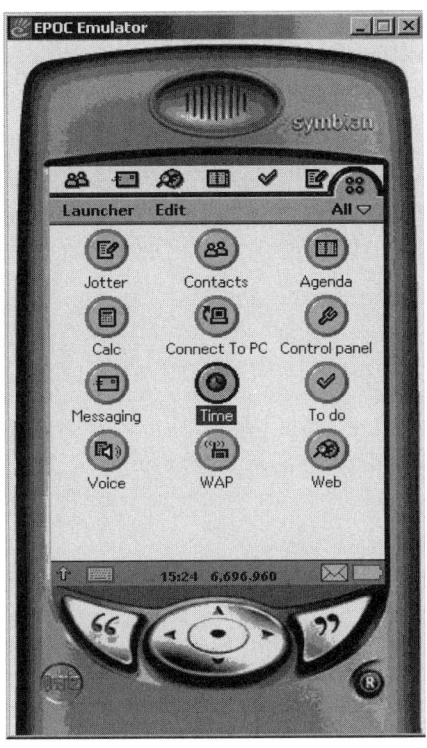 

Figure 7.16 *EPOC Emulator.*

printing, e-mail, contact and calendar synchronization, and even application installation.

7.3 Management

No application can be deployed on a wide scale without management support. In the case of mobile devices, which are frequently off site, this task can be very challenging.

Without a good network connection it is not easy to monitor, troubleshoot, and fix common user problems. Remote control software, if available for that platform, can simulate physical access, but it too has shortcomings when it is necessary to connect peripherals or press physical buttons.

A management solution also needs to ensure that an effective backup/restore solution is in place. And a good tracking system is necessary if an accurate inventory of mobile units is required.

But probably the most challenging management task is the installation and update of software on the devices. Even when the set of devices is very uniform, tests need to be done to determine the impact on a very limited footprint.

Downloading applications can be very slow over a wireless connection. But a recall of the devices accompanied by a manual installation also has its disadvantages.

7.4 Summary

The design and implementation of an actual wireless solution begins with the selection of the individual components, such as mobile devices, air interfaces and network providers, and wireless gateways. These are typically dictated by the needs of the users in terms of which applications they require when and where.

In the future, you can increasingly expect applications and devices to be aware of the mobile paradigm and wireless (or at least unreliable and slower) networks. It should be possible to run any application over any network with any device. However, this is not yet a reality.

To fill this interoperability gap a number of vendors have developed "middleware" solutions that facilitate the use of legacy applications with modern devices. Unfortunately there is very little commonality between these offerings. It is a laborious task to compare the many products available with the device and application requirements of a particular company or department. Ultimately though it is worth the effort to ensure that the end-user is able to use wireless technology in a productive manner.

After all the components have been selected comes the network design both in terms of a logical topology and also the physical placement of all the servers. Usually wireless networks are built as an extension to existing infrastructure so the basic design is dependent on the networks in place. The physical placement and sizing of the servers is important to ensure high-availability and acceptable performance. This may involve load-balancing and decentralizing servers for optimized local access.

In many cases off-the-shelf applications will not completely fill all user requirements. When it comes to in-house or contracted development of ad-hoc applications it is important to ensure that they fit the wireless model from the start. They should be mobile-aware in that they are able to cope with poor networks, can adapt the user interface to the machine characteristics of the client, and are able to operate in disconnected mode with subsequent synchronization.

One final challenge in implementing wireless solutions is to ensure that they are easily and professionally managed. The hardest part to this is the deployment, updating, monitoring and support of mobile devices. Not only the diversity of devices but also the fact that they are often remote from the IT department makes this task much more complex than traditional desktop support.

Bibliography and related Web sites

Client platforms

Murray, J. *Inside Microsoft Windows CE*. Microsoft Press, 1998.

Rhodes, N., and J. McKeehan. *Palm Programming*. O'Reilly, 1999.

Tasker, M. *Professional Symbian Programming*. Wrox Press, 2000.

Microsoft Windows CE factsheet:
http://www.microsoft.com/windows/embedded/ce/tools/factsheet.asp

Middleware

Microsoft Mobile Information Server: http://www.microsoft.com/miserver/

Extended Systems: http://www.extendedsystems.com/

Infowave: http://www.infowave.com

724 Solutions: http://www.724.com

Brience: http://www.brience.com

Aether: http://www.aethersystems.com

IBM WebSphere: http://www.ibm.com/websphere

AvantGo: http://www.avantgo.com

ViaFone: http://www.viafone.com

8

Future Directions

In this chapter, we will look at some of the current developments in wireless technology. At this point it is not possible to know where they will lead, but it is clear that some of them will eventually make it into mainstream technology.

We will begin with the transformation of mobile architectures based on IP-based networking. By providing a common denominator for all wireless technologies (and, in fact, one that is also common to wired technologies) the architecture becomes more open and enables rapid development of independent modules.

We will also look at the possibilities for improving air interfaces to maximize throughput and versatility, as well as the evolution of current wireless networks to improve service and mobility.

New applications will need to be written with mobile and wireless environments in mind. We will examine some of the implications for software developers and the evolution of tools to support them. At the same time, the higher bandwidth and quality of service will enable a new set of applications, such as streaming media.

8.1 Mobile architectures

8.1.1 IP-Based networks

One trend that some mobile networks have already begun is the migration to IP-based networks. Wireless LANs almost all use IP, and WANs, such as GPRS, are beginning to follow suit.

Note that the common denominator is IP as the routing protocol. There will be transport protocols other than TCP. In mobile environments UDP is

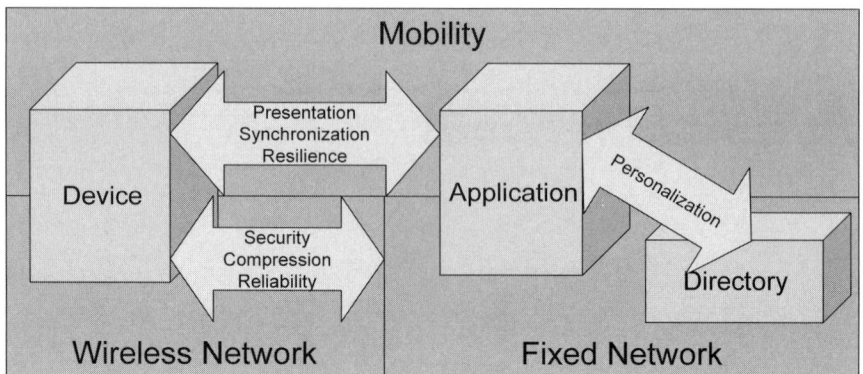

Figure 8.1
Wireless and fixed networks.

often more appropriate. Telecom companies are also pursuing Stream Control Transmission Protocol (SCTP) as a transport protocol tuned to signaling. But all of these are end-to-end protocols. At the networking layer both wireless and fixed networks simply route IP traffic. (See Figure 8.1.)

It is also important to distinguish IP from the Media Access Control (MAC) and physical (PHY) layers. Wireless technology clearly uses media different from copper and optical fiber. There will always be a need to tune link-level protocols to the media they support. The trend to note in this context is that IP is becoming available over all these physical layers.

This convergence allows the network to operate more efficiently, since it is streamlined for a single protocol. But it also offers a major advantage to end devices and applications. They can easily add functionality since, even though we have a very heterogeneous set of scenarios, we have one common denominator: IP.

This simplification allows network technology to focus on its primary purpose. The job of network technology is to provide a secure, reliable, and efficient link from the device to the application server. It doesn't need to do anything specific to mobile but needs to cover the weaknesses in wireless (latency, packet loss, low bandwidth, shared medium) so that applications can use wireless transparently.

8.1.2 Mobile-aware applications

Applications don't need to know about wireless per se, but they do need to know about mobile devices and slow/unreliable networks.

There are other slow and unreliable networks (such as dial-up, multihop, and international connections). Applications need to be resilient to these.

They cannot fall over when a packet is lost or time out quickly on a connection. Whether this is covered within the application logic or the networking modules in the platform, there needs to be resilience.

Applications also need to be aware that devices can offer many different form factors and machine interfaces. This means they must either have an adaptable user interface or else provide device-dependent versions.

They must also be able to cope with the fact that the device will not always be connected. The user should still be able to continue working at least to a limited extent with cache or synchronized data.

8.1.3 Open mobile platforms

In order to provide full off-line capabilities it is necessary to have client applications. This will be much simpler with open mobile platforms, where any software developer can put together a client application and install it on the device. Older, proprietary platforms will not be able to keep pace with rapidly changing needs of mobile users.

8.1.4 Universal repositories of personal information

Applications need to have access to a large amount of context information in order to be able to provide an efficient user experience. Historically directories (typically LDAP directories such as Microsoft Active Directory) have stored user information and made it available to applications.

There are two limitations to current directories. They typically store only limited information and they are not universally available.

Future directories need to access traditional personal information, such as phone number, address, and full name, but they also need more transient information, such as presence, current device type, location, and current IP address.

They also need to be universally accessible. The Microsoft Hailstorm initiative is one attempt to create a worldwide integrated repository of all personal information. These data are exposed as a Web service, which and is available to any commercial site or corporate system. The effort is an ambitious attempt to expand on the success of Microsoft Passport, with uniform authentication while also alleviating the user from duplicate entry of context information.

There are many public concerns surrounding Hailstorm, not the least of which is whether users want to entrust all their personal data to Microsoft.

Whether the initiative will prevail is unclear. There may be other competing offerings, or governments or independent international agencies that will eventually provide similar infrastructure.

In any case, we can expect a much more active role of directories in mobile environments in the future.

8.2 Air interfaces

8.2.1 Third-generation networks

Before we dive into third-generation wide area cellular networks, let's take a step back and look at what 3G really means.

Table 8.1 describes the features of the first-generation (1G), second-generation (2G), and third-generation (3G) mobile phone systems. There is also an interim step, usually just called 2.5G.

While there is a general consensus that the first generation systems were analog implementations, such as AMPS, TACS, and NMT, there are at least two different interpretations of the following generations.

The first approach is to look at the bandwidth. Today's second-generation systems typically provide approximately 10 Kbps. Third-generation systems should offer on the order of 1 Mbps, and the 2.5G systems offer approximately 100 Kbps. While inexact, this does give a broad picture of the evolving data rates.

There is another tendency in this evolution though. Today's systems are based on voice and therefore are largely circuit switched. Eventually everything will be based on IP and therefore packet switched. But in the interim, most 2.5G systems will offer packet-switched data while still maintaining circuits for voice traffic.

Table 8.1 *1G, 2G, 2.5G, and 3G Mobile Phone Systems*

Generation	1G	2G	2.5G	3G
Bandwidth		~10 Kbps	~100 Kbps	~1 Mbps
Voice Traffic	Circuit switched	Circuit switched	Circuit switched	Packet Switched (VoIP)
Data traffic	No data	Circuit switched	Packet Switched	Packet Switched
Modulation	Analog	Digital		

8.2 Air interfaces

Most technologies fit into the same generation using either set of criteria, so there is little controversy surrounding the generations. However, the match is not completely solid. Purists may ask themselves where to place HSCSD with 57 Kbps of circuit-switched data or Japan's PDC with 9.6 Kbps of packet-switched data.

Why not implement 3G today?

It would seem much more efficient to bypass some of these intermediate standards and go straight to the third-generation networks. However, the process is much longer and more complex than one might expect. Once the standards have been completely ratified, the telco equipment manufacturers must develop and extensively test their base stations and accompanying infrastructure.

The handset manufacturers also need to produce and miniaturize devices tuned to the specification. It is a major challenge to balance the conflicting requirements of processing power, range, reliability, and battery lifetime. All of these components then need to be tested for interoperability with the infrastructure, which must be deployed over extensive geographic areas.

Going through all of these steps can take years, especially since it is very common to encounter obstacles that had not been predicted in early planning. It therefore makes sense to pursue cheaper and quicker upgrades to the network that provide only limited improvements. But at the same time, the third-generation project continues at its own pace, confident to be the long-term victor.

GSM evolution

We already mentioned the migration path of GSM in Chapter 3. As discussed, it incorporates two trends: increased data throughput (HSCSD—56 Kbps, GPRS—171 Kbps, EDGE—385 Kbps, UMTS—2 Mbps) and the migration from circuit-switched data to packet-switched data.

High-Speed Circuit-Switched Data (HSCSD) offers better compression (14.4 Kbps per slot) than CSD (9.6 Kbps). It can use up to four slots (4 × 14.4 = 57.6 Kbps), although two slots is more common (28.8 Kbps). Its implementation requires both an HSCSD-enabled client and support in the infrastructure. Network upgrades are relatively simple and cheap, but many operators have ignored it, as they prefer to focus on GPRS and UMTS.

The mobile networks started to roll out GPRS in 2000. Most European operators now claim to offer some level of GPRS service. In theory it was a "cheap" upgrade from GSM, but, by the middle of 2001, quality was still poor, service was unstable, and many users complained about interference between voice and data traffic. All GPRS devices have been set back in their schedule, but slowly they are now appearing on the market. There is little immediate impact on the applications, other than improved speed. The real benefits will be reaped when applications are developed to take advantage of the always-on nature of GPRS to provide instant notification.

Enhanced Datarate GSM Evolution (EDGE) is really a poor man's 3G. Many operators either did not bid or were unsuccessful in acquiring UMTS spectrum. They now have the option to purse EDGE in the existing GSM frequency ranges. In many ways it is similar to GPRS. The main difference is modulation, since it uses eight-shift keying instead of GMSK and can therefore achieve higher data rates.

Unfortunately, the modulation change implies a replacement of most of the mobile infrastructure, which is a big obstacle for most GSM operators who will focus on GPRS and then UMTS. On the other hand IS-136 operators may decide to make EDGE rather than GPRS their first step, since they will need to replace their base stations either way.

UMTS is represented with two interfaces: W-CDMA and TD-CDMA, both in the 1,900–2,200 MHz range. Its specifications stipulate bandwidth requirements as follows:

- 384 Kbps—full area coverage
- 144 Kbps—moving vehicle (120–500 km/h)
- 2 Mbps—local coverage

In contrast to earlier proposals, the most recent revision (UMTS R'00) requires all transmissions to use IP and, in particular, IPv6. Even voice traffic must be run over IP using the SIP protocol.

Table 8.2 describes the differences between the two UMTS interfaces. W-CDMA was proposed by Ericsson and Nokia and is allocated in the paired bands, which usually occupy spectrum in the 1,920–1,980 MHz and 2,110–2,170 MHz ranges. The paired bands allow the up link and down link to use different frequencies, a duplexing technique known as Frequency Division Duplexing (FDD). This is ideal for symmetric traffic, such as a typical voice conversation.

TD-CDMA was proposed by Alcatel, Siemens, and Nortel and uses the unpaired bands (typically 1,900–1,920 MHz and 2,010–2,025 MHz—

Table 8.2 *W-CDMA and TD-CDMA Differences*

	UMTS 1,900–2,200 MHz	
Air interface	W-CDMA (Wideband CDMA)	TD-CDMA (Time-division CDMA)
Proponents	Ericsson, Nokia	Alcatel, Siemens, Nortel
Spectrum allocation	Paired bands	Unpaired bands
Duplexing	FDD (Frequency division duplexing)	TDD (Time division duplexing)
Tuned for	Symmetric traffic	Asymmetric traffic

whereby 2,010–2,020 MHz is unlicensed). Since both the up link and the down link use the same frequencies, they must alternate time slots (which is called TDD—Time-Division Duplexing). This is much more efficient for asymmetric traffic (such as bulk downloads).

Asia is heading the charge of the UMTS roll-out. In Japan, trials are currently in progress with plans for commercial availability in 2002. They have more incentive since there is no "real" (high-speed) 2.5G network in place. It is important to realize that UMTS deployments are linked to IPv6 roll-outs, which are also more advanced in Asia than elsewhere. Meanwhile, many European operators are broke after the spectrum auctions and have uncertain roll-out dates. Already heavily indebted they are now conservative about investing too much into 3G testing and deployment without solid projections on their return on investment. In order to ensure earlier profitability some, such as Finland's Sonera and Sweden's Telia, are hedging their bets with WLANs.

3G alternatives

Having more than one international standard is awkward for everyone. It stifles the market and reduces the capability of global roaming. For over a decade the International Telecommunications Union (ITU) has attempted to rationalize the standards into one interface. At one point it had to review over a dozen competing standards for 3G technologies. Two of the strongest contenders were Qualcomm and the GSM association.

Qualcomm wanted a standard (cdma2000) that was backward compatible with IS-95 to provide cheap upgrades for its customer base. However, it also demanded high patent royalties for the technology. The GSM association balked at the fees and instead selected another technology (UMTS) using less (but still some) of Qualcomm's intellectual property.

While some countries (China, in particular) have raised the option of developing other 3G standards, it now appears likely that cdma2000 and UMTS will be the only two terrestrial wide area networks to gain widespread acceptance.

The standardization efforts of these two technologies are now coordinated within two projects, called the Third-Generation Partnership Project. These two technologies are as follows:

- 3GPP 1: specifications for GSM (and IS-136 and PDC) evolution to 3G

- 3GPP 2: specifications for CDMA (IS-95) evolution to 3G

As with anything new, it remains to be seen if one or both of these technologies can establish themselves long term. UMTS has been selected as the air interface by the vast majority of the world's mobile operators. Not only those with a GSM heritage but also IS-136 and PDC operators have selected UMTS. In Korea there are even IS-95 CDMA operators who have stated their intention to implement UMTS in order to maintain compatibility with the rest of the world.

IS-95 operators who elect to go to cdma2000 do have one major advantage. They have a relatively cheap upgrade path, while UMTS requires very expensive replacements in their infrastructure. However, both of these newer technologies may become obsolete before they become commercially viable. WLAN provides better data transfer at typically cheaper costs, and users may find that a combination of WLAN for stationary data in metropolitan areas complemented with 2.5G for wide range data access and voice capabilities is sufficient.

8.2.2 What about 4G?

Nobody expects 3G networks to be the end of the road. Wireless networks will continue to evolve. But what should we classify as 4G? Using the criteria mentioned earlier, one approach would be to use the term to designate very high bandwidth (10 Mbps+) required for streaming media. This is difficult to achieve on a wide range using current transmission technologies, but can it be accomplished with new transmission technologies?

Alternatively, is it a combination of overlaid networks? Imagine a scenario where the same device can use Bluetooth, 802.11a, 802.11b, 2.5G, 3G, high-bandwidth 4G, and even satellite transmission, depending on coverage, bandwidth requirements, and service pricing. If, in real time, it can switch between the technologies to transparently provide the best and cheapest service available, this will clearly be a major improvement in functionality, enough to warrant its own generation.

8.2.3 Transmission technologies

OFDM—Orthogonal Frequency Division Multiplexing

Spread spectrum uses different carrier waves to send signals. But it only uses one at any given point in time. Multicarrier Modulation (MCM), and its implementation with OFDM, goes one step further by simultaneously transmitting at many frequencies.

In the past this would have been difficult and expensive, since each carrier wave would have required a separate oscillator. However, it is now possible to use Digital Signal Processors (DSPs), which can superimpose the carrier waves using a mathematical technique called Fast Fourier Transform (FFT). Only the resultant combination needs to be transmitted by the radio.

The result is increased spectral efficiency, better resistance to multi-path interference, and improved performance in non-line-of-sight environments. There have been several different implementations of OFDM, including Vector OFDM (VOFDM), Wideband OFDM (WOFDM), and Flash-OFDM, but all use similar techniques.

Initially, OFDM was barred from the 2.4 GHz spectrum, since OFDM was not considered spread spectrum. This was a controversial decision, since the two techniques do have some commonality. More recently the U.S. FCC has reassessed this decision. However, at present, most OFDM is targeted to other frequency ranges. Both HiperLAN2 and 802.1a will use OFDM. The IEEE 802.16 Working Group for Wireless MANs also proposes OFDM for use in the wireless local loop. At present the 3G WAN networks do not use OFDM.

Smart antennae

Most antennae are omnidirectional. They transmit in all directions with similar intensity, as shown in the following graphic.

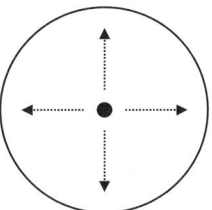

However, there are also directional antennae, which transmit primarily in one particular direction, as shown in the following graphic.

Adaptive (or smart) antennae are able to vary the direction in order to maximize performance. For example, some antennae can determine the direction of arrival of incoming signals and can respond in the same direction, as shown in the following graphic.

Smart antennae are able to transmit different signals in different directions at the same time. They can achieve better range and coverage and increase their capacity. They are also less prone to multi-path interference.

This resembles switched data. Although it is insufficient as a security mechanism, there is a certain degree of privacy inherent in directional transmissions. By monitoring the direction of a given user it is also possible to determine the position with much more accuracy than using omnidirectional antennae.

Software radios

Early radios contained a set of hard-wired components. The oscillator used a fixed frequency and the signal was modulated on top of that. We have already seen how a Digital Signal Processor (DSP) has expanded the possibilities of a radio enormously and made it possible to achieve much better throughput and higher reliability by responding to the immediate environment.

The next step in the process is to build radios that dynamically adapt to whichever wireless technology is most suitable. As has been evident throughout this book, there are many different wireless technologies using a wide variety of modulation techniques and frequencies. And these technologies are all evolving at a staggering pace.

It would not be effective to build radios that can utilize all these transmission techniques. Putting multiple radios on the same device is out of the question for power and space reasons, not to mention the possibility of mutual interference.

What does make sense is to control the DSP through software. It is easy to update the software with new versions, and the software can make intelligent choices as to which transmission technique(s) to utilize at any given time. With modern technology it can even superimpose two or more signals on one transmission.

The use of a software-controlled radio is obvious. There is no limit to the amount of processing that it could potentially embed. The term "cognitive radio" has even been coined to refer to devices that are capable of representing complex environmental knowledge, derived from the location, and therefore able to tune the radio to the varying needs of a mobile user.

8.3 Networks

8.3.1 QoS—Quality of Service

The transition from circuit-switched to packet-switched data provides increased efficiency of the network and higher overall throughput. But it is not without its disadvantages. Many applications (such as VoIP or streaming video) require a very reliable low-latency transport. Packet-switched networks operate on a best-effort basis and cannot guarantee this service, particularly when the network load is high.

Quality of Service (QoS) is an attempt to reintroduce reliable guaranteed bandwidth into a packet-switched network. There are different approaches to this that vary in terms of stringency as well as the networks for which they are suited. The two main techniques used are resource reservation (which guarantees a fixed channel for the duration of the session) and prioritization (which merely provides a means for designating that some traffic flows are more sensitive to latency than others).

At the IP level there are two common protocols used to achieve these goals. ReSource Reservation Protocol (RSVP) provides a means of reserving bandwidth at each node along a specified path between two entities. It is

Table 8.3 *Real-Time and Best-Effort Traffic Classes*

Traffic class		Characteristics	Description
Real time	Conversational	Stringent low latency	Voice, video conferencing
	Streaming	Preserve time relation (constant latency)	Streaming audio, video
Best effort	Interactive	Medium latency and bandwidth	Web browsing, database lookup
	Background	Preserve payload content	File transfer, e-mail download

usually implemented in conjunction with Integrated Services (IntServ), providing at least two levels of service: Guaranteed, which emulates a virtual circuit and effectively guarantees immediate delivery of data, and controlled load, which simulates a controlled load; it cannot guarantee a specific service but does provide much better delivery time than default best effort.

DiffServ uses simple prioritization. It defines two services levels (usually called PHBs—per-hop behaviors). Expedited forwarding assures that within bounds all bandwidth is forwarded with very little latency and the rest is discarded. Assured forwarding provides many different control points, which can be used to define more complex rules of prioritization.

Some other protocols that are related to QoS include Multiprotocol Label Switching (MPLS), which provides bandwidth management via network routing control using packet header labels. It allows differentiated routing depending on the packet type. Subnet Bandwidth Management (SBM) and 802.1p provide QoS at the data-link layer on 802 networks.

UMTS QoS classes

Since one of the goals of UMTS is to provide connectivity for multimedia applications, it relies very heavily on QoS. As shown in Table 8.3, UMTS defines four QoS traffic classes, which map directly to the IPv6 traffic_class field and permit the application to request expedited services. Clearly, the more demanding the QoS requirement from the application, the higher the billing rate will be for that data flow.

8.3.2 Mobile IP

The increasing requirement for mobility is difficult to balance with the design of the Internet, which assumed static connections. In particular, IPv4, the version currently in use throughout the world, is incompatible with the goal of seamless roaming between networks.

8.3 Networks

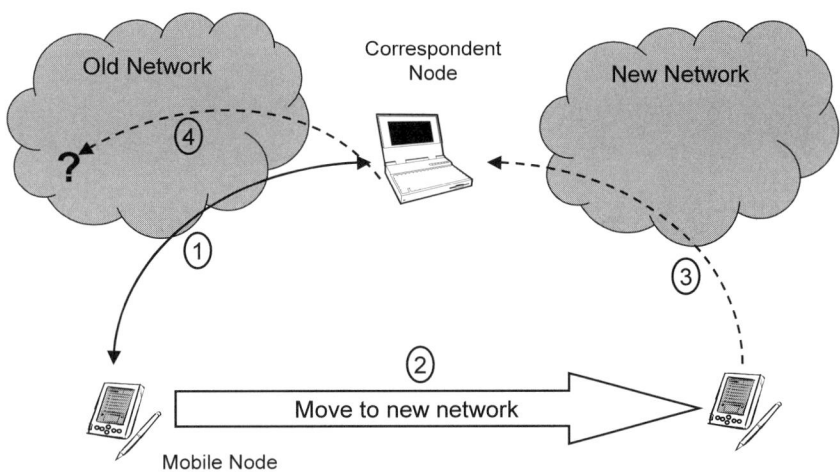

Figure 8.2
Moving a system to a different network.

The crux of the problem is the assumption of IPv4 that any given network session will use static IP addresses and ports. Over the life of a single interaction the following quadruple will remain constant.

> Source IP Address   Source Port Number
>
> Destination IP address   Destination port number

The problem with this assumption is that moving a system to a different network (on another IP subnet) inevitably changes the IP address.

Figure 8.2 portrays the dilemma, which is characterized as follows:

1. Two devices begin a session through a network. They are both able to send each other data at their static IP addresses.
2. The Mobile Node (MN) physically moves and is therefore connected to another network with another subnet. During this process it changes IP address.
3. The MN is still able to send data to the Correspondent Node (CN), since the address of the CN has not changed.
4. However, the CN will still try to send information to the MN using the old IP address. These packets will be discarded, since the old IP address is no longer in use.

In order to maintain the link it is necessary to install some additional infrastructure to ensure that all traffic to the MN is rerouted from the old network to the new network.

Figure 8.3
Every mobile node is associated with a home network.

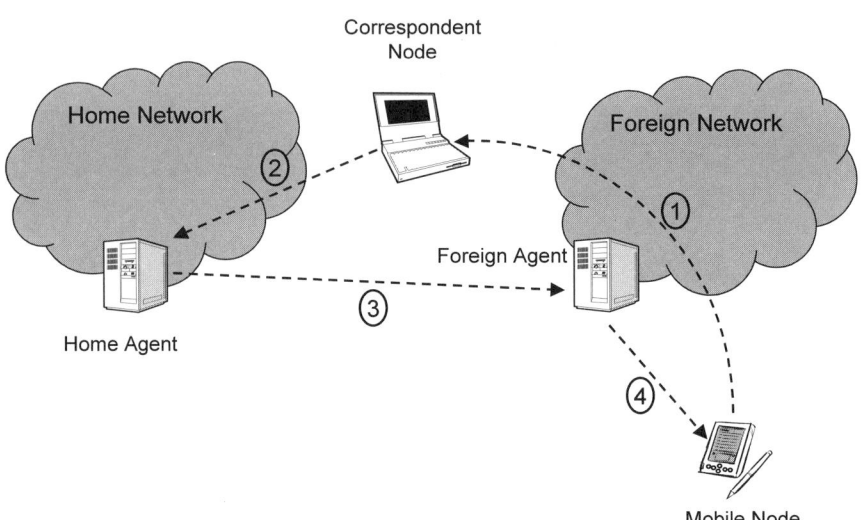

As shown in Figure 8.3, every Mobile Node (MN) is associated with a Home Network (HN). On every home network there is a Home Agent (HA) that maintains location information about the mobile node. When the MN roams to a foreign network, it must register its location (in the form of a care-of-address) with the HA.

The result is that the mobile node can continue its sessions uninterrupted, as follows:

1. The data traffic from the mobile node to the correspondent node is not affected.
2. But when the CN sends data to the old home address, the home agent intercepts the packets.
3. The HA then tunnels these packets to the foreign agent on the foreign network.
4. The foreign agent unpacks (detunnels) the packets and sends them on to the mobile node.

This scenario will allow mobility with IPv4 without requiring any changes to all potential correspondent nodes. However, it is obvious that the process is far from efficient. Rather than using a direct path from the CN to the MN, all data must pass through both a home agent and a foreign agent on two potentially distant networks (triangular routing).

The result is higher network utilization, a heavier load on the agents, and greater latency on all the traffic from the CN to the MN. It is possible to optimize the route; however, this does require that all correspondents be modified.

IPv4 has been widely deployed without Mobile IP. Although it is possible to retrofit it, as described previously, the result is still less than ideal. Additional infrastructure is needed in terms of home and foreign agents on all subnets. And even so, the traffic is usually inefficient since it involves triangular routing.

8.3.3 IPv6

Predictions of the global opportunity of past networking standards, including OSI networks, have not always materialized. IPv6 has been hailed as the imminent savior of the Internet since the early 1990s. And yet IPv4 has managed to maintain its position as the dominant routing protocol, accommodating any inadequacies with supplementary standards. It has been able to keep working reliably and relatively efficiently for long enough that many skeptics question whether there will ever be a need to replace it.

The most visible weakness of IPv4 is its limited address space. With only 2^{32} potential addresses it is not able to provide a unique number to every device connected to the Internet. The interim solution to this problem has been Network Address Translation (NAT), which can segment the network into an almost unlimited number of duplicate private address spaces. However, this technique is not without its shortcomings, since it introduces other problems, not the least of which is an increased burden on the routers performing NAT.

In addition to a much larger address space (2^{128} addresses), IPv6 includes several enhancements that are indispensable to modern networks. Some of these include Mobile IP, Quality of Service (QoS), and IP security (IPsec).

Mobile IP provides a means for a device to move from one subnet to another (and thereby change IP address) while maintaining an active connection. This implies a need for the home network to be aware of the current location of the device and to forward all traffic to the destination network.

All IPv6 networks must support Mobile IP, and they are able to route traffic directly to the destination without passing through the home network.

An MN acquires its care of address through normal IPv6 Address Autoconfiguration and Neighbor Discovery. It does not use a foreign agent.

Instead, route optimization is built in. The MN sends an address update to all Correspondent Nodes. Packets may then be sent directly to Care-of Address without passing through an agent.

Mobile IPv6 is therefore much more efficient than Mobile IPv4 in routing traffic, since it performs much less redirection and instead is able to bind directly between end nodes. This also implies less infrastructural support. A foreign agent is not needed at all and the home agent is either not used at all or else only used once to find the current address of the mobile node.

The attempt to replace circuit-switched connections with packet data is exposing another deficiency of IPv4. Voice over IP and other streaming services require dedicated bandwidth in order to operate reliably. There are mechanisms to indicate quality of service in IPv4 but, again, IPv6 contains this feature natively.

IPv6 header has two QoS-related fields, as follows:

- 20-bit flow label, which is primarily geared to IntServ, but may have other uses
- 8-bit traffic class indicator, which is usable with (but not restricted to) DiffServ

IPv6 does not automatically provide better quality of service than IPv4, but the fact that it is a more efficient protocol allows it to scale upward much more easily.

The emergence and widespread use of ISPs has produced an enormous range of security issues. To address these, it has become necessary to ensure some kind of security over public networks. IPsec is the most promising security technology over IP networks, both for point-to-point connections and tunneling between private networks.

IPv6 stipulates IPsec as a security mechanism, which will ensure that it is globally implemented. IPv4 can also use Ipsec, but it is not universally adopted and is more difficult to implement. In particular, it is important to note the IPsec is incompatible with NAT; therefore, it is not feasible for use over many IPv4 networks.

At the same time many mobile operators (who will soon be providing IP connectivity) and traditional ISPs have more users than can be accommodated in one Class A subnet (2^{24} = 16 million). This leads to workarounds, such as internal Network Address Translation, that are both inefficient and restrictive. For example, they limit the ability to place a server on the network.

The squeeze is felt most strongly in Asia, where huge populations are vying for a very small number of allocated addresses. In China and Japan, government officials have already announced that public ISPs must migrate to IPv6 by the middle of the decade.

Mobile operators are also unable to obtain the large address ranges they need as they transition to third-generation (3G) IP-based networks. The Third-Generation Partnership Project (3GPP) recognized this obstacle and has now mandated that all UMTS implementations will require IPv6.

Among the other advantages of IPv6 are its efficiency at routing and management, particularly in the context of large and complex networks. It includes simple mechanisms for autoconfiguring and renumbering addresses, which makes the network much more flexible to accommodate changes, a frequent need in today's dynamic environment.

8.3.4 Migration to IPv6

In order for an IPv4-only system to communicate with an IPv6-only system (and vice versa) it is possible to use a Protocol Translator (PT). The PT is placed between the systems and translates the lower-level IP traffic from one version to the other while leaving the upper-level protocols intact.

An alternative to protocol translation is to run a dual stack on at least one of the end systems. This will probably be the more common approach. The first step is to begin installing IPv6 stacks on the current devices and back-end systems while continuing to run the applications over the IPv4 stack. Once a critical mass of systems support IPv6 it is possible to begin transitioning sessions to the new stack. However, both stacks must remain until all connections are running over IPv6, a process that is likely to take many years in large enterprises.

The migrations of the applications to IPv6 should be straightforward. In theory, the network layers (i.e., IPv4 versus IPv6) should be abstracted from the application by the sockets interface (except for base infrastructure applications—DNS, DHCP, RAS, PPTP, and others). However, applications that have manually extracted IP addresses into fixed variable lengths may encounter issues reminiscent of Y2K and will need to be adapted.

In addition to migrating the end systems to IPv6 it is necessary to ensure that connectivity is possible at all hops along the path. In many cases the

IPv6 networks may not be directly connected and will form islands within the IPv4 ocean. There are two ways of interconnecting these isolated networks: 6to4 and 6over4.

The more common approach is likely to be 6to4. All IPv6 traffic is tunneled through IPv4. The edge routers must be dual stack and are the only components involved in the tunneling process. It is transparent to the end systems. An alternative mechanism for connecting to IPv4 routers that are not IPv6 aware is 6over4. However, this technique is not expected to be used frequently.

8.4 Presentation

8.4.1 Convergence of markup languages

There are a number of markup languages in use today. From traditional HTML to WML to i-mode's cHTML, HDML is still supported by OpenWave browsers and there are more. The World Wide Web Consortium has selected a markup language called XHTML, which is XML based and will be the migration path for HTML.

In an effort to converge with the efforts of the W3C, the WAP Forum has also decided that WML 2.0 will effectively merge with XHTML. It is likely that others will follow suit and take advantage of the market endorsement of XHTML.

It might seem like this will solve all the problems of presenting content to mobile devices in a uniform manner. Unfortunately, it is not that simple. While it is certainly a big advantage if there is a single syntax to describe all Web content, the need to provide it in a format that is suitable to a wide range of devices remains.

The real challenge is in tailoring the content to the machine interface of the device. It is not in translating it from one syntax to another.

Future of WAP without WML

If WML converges with XHTML, will there be any need to retain the WAP at all? As we move to 3G technologies most of the motivation for a secondary stack is dissipating. One might be tempted to ask what the WAP Forum will work on next.

The answer is that the WAP Forum is working on an amazing number of new standards and protocols independent of the WML and the WAP stack. The WAP Push Architecture defines two protocols: Push Access Protocol (PAP) and Push Over-the-Air Protocol, which provide a common means of automatically updating a mobile device from the central server.

There are initiatives involving graphics and pictograms for restricted machine interfaces and multimedia messaging services for all mobile platforms. The WAP Forum is very active in the area of security, from content signing to smart card support and SIM integration.

They are also active in a wide variety of other initiatives, including data synchronization (SyncML), billing frameworks, caching models, and location frameworks. For the foreseeable future there is enough uncharted territory in the area of wireless standardization to keep the WAP Forum busy.

8.5 Development tools

With the realization that a high percentage of tomorrow's clients will be running on mobile devices comes the necessity for new applications to take this reality into consideration from the start.

Whether the application is client-based or server-based, an application developer should be able to write an application once for all devices. If it is client-based, it needs to be transparently portable and automatically adapt to the native user interface of the device. If it is a server-based application, it needs to render to the client automatically at run time.

Some tools have already begun to emerge to address this need (such as Microsoft Mobile Internet Toolkit or Salsa Systems), but the task is large and complex so it will take some time before the full range of devices (including Palm, Windows CE, EPOC, and all relevant markup languages) is supported. Good simulation capabilities must complement the development environment to ensure proper and thorough testing. As multi-modal machine interfaces become the norm, the application development tools need to support multiple simultaneous modalities automatically.

But there is more to making a mobile application than tailoring the user interface. Since mobile devices will always have the need to disconnect, it is important that the application be able to synchronize and cache information about the client for offline use. Whether these devices use SyncML or another synchronization technique, the capability should be an integral part of the development environment so that application developers will easily be able to provide a consistent means of synchronization.

8.6 Applications

8.6.1 Voice over IP

In the past few years, the interest in Voice over IP has surged. It represents an alternative to traditional switched telephony that can take advantage of the now ubiquitous Internet. Its benefits include the following:

- Consolidation—reduced costs of infrastructure (including reduced operation and administrative costs), since there is no duplication of voice and data infrastructure. Everything runs on one single packet-switched network.

- Lower costs for WAN calls—the bandwidth between voice and data is shared; since it is already in place, there is no need for use-based charges. Through compression, it results in more effectively utilized network resources (keep in mind that the telephone networks allocate 64 Kbps per channel even for silence on both ends).

- Increased user mobility—although modern telephony has made great strides in accommodating mobility, IP will be able to immediately reroute traffic to a user no matter where he or she is on the network.

- Integration of services—as illustrated with collaboration tools such as Microsoft Netmeeting or Groove, voice, video, remote control, and file transfer can all use a single channel.

The challenge of Internet telephony is to provide at least the same level of voice quality as the traditional telephone system. This means ensuring clarity of transmission while at the same time minimizing echo and delay.

To support this, VoIP requires a good level of underlying QoS. To some extent the quality is dependent on the Codecs being used. Those that compress more require less bandwidth, but they also result in less clarity. There is also considerable variation in their sensitivity to latency and packet loss.

At present, three VoIP protocols have gained significant market interest: H.323, SIP, and MeGaCo.

H.323

H.323 is an ITU standard designed for multimedia over LAN and is often used in conjunction with the T.120 series recommendations that provide collaborative features, including interactive gaming, virtual reality, simulations, and file transfer. H.323 is very complex for voice alone. However, it is the most mature and therefore the most common protocol in use today.

SIP

Session Initiation Protocol (SIP) is an IETF standard. It is more consistent with other Internet protocols and uses a syntax similar to SMTP and HTTP. It assumes QoS is in network (e.g., in the form of RSVP, DiffServe) and does not provide any itself. Due to its simplicity SIP is considered a favorite in the emerging VoIP market. For example, it has been selected by the 3GPP for use in UMTS.

MeGaCo

MeGaCo was designed to emulate the public telephone network. It can work with any common phone and is therefore cheaper to implement. This is in contrast to H.323 and SIP, which both require intelligent end points, such as PCs.

Since working with networking protocols is an unnecessary level of detail for most programmers, there are several telephone APIs that can be configured to use any of the protocols. These include the following:

TAPI (Telephony API)—provided by Microsoft for Windows platforms

TSAPI (Telephony Services API)—provided by Novell for NetWare environments

JTAPI (Java Telephony API)—provided by Sun for Java applications

Other than providing a decent quality level, the biggest challenge for VoIP is interoperability with the PSTN. There are some fundamental differences in the approaches that must be bridged. At some point it is necessary to translate between IP addresses usually stored in DNS and the telephone numbers (geographically mapped according to ITU E.164).

A gateway must also make the actual connection between a public switched telephone and a VoIP terminal. Ensuring that there are sufficient gateway resources and that the optimal gateway is selected (there can be considerable variation in long-distance phone charges depending on the location of the gateway) is not a trivial exercise.

VoIP also faces the need to incorporate many of the modern advances in telephone services, including the following:

- Call blocking/trace/waiting/forwarding
- Caller ID
- Hold, transfer, conference

- Directory services
- Encryption

Since VoIP is already digitized, the challenge of implementing these is smaller than it was for previous analog telephone systems. However, this does not mean that all VoIP systems offer them today.

High bandwidth

We've already mentioned that one of the driving factors for 3G is the growth of high-bandwidth Internet applications. There is an increasing need to support software such as the following:

- Audio (including MP3)
- Still and moving images
- VoIP—Voice over Internet Protocol
- Picture messaging
- Passive video (video and audio on demand)
- Teleworking
- Telemedicine (monitoring patients at home, ambulance)

3G networks will be improving QoS to support these applications and will be increasing the bandwidth available. However, this does not mean that everything will be available with 3G.

From the bandwidth requirements shown in Table 8.4, it is obvious that even in favorable circumstances UMTS will not be able to provide video with the quality and resolution of HDTV.

8.6.2 Multicasting

Wireless is primarily used for point-to-point connections. However, since it is a shared medium, it is actually very suitable for point-to-multipoint sessions.

Television and radio broadcasting technologies use wireless for their transmissions. To date this has been disjoint from wireless data transfers. But as the bandwidth and reliability of our networks increase, it makes sense to investigate distributing content via the Internet—that is, over IP-based wireless networks.

The advantages are twofold. Broadcasters are able to reach the entire world without needing to have a radio transmitter. This means less infra-

Table 8.4 *Bandwidth Requirements*

Application	Bandwidth
Telephone	
High quality	34–64 Kbps
Low quality	4–11 Kbps
Audio	
MP3	128 Kbps
Dolby digital	768 Kbps
Redbook	1.45 Mbps
Video	
MPEG-1	3 Mbps
MPEG-2 (DVD, DVR)	4–11 Mbps
MPEG-4	1–2 Mbps
HDTV	20 Mbps

structural and operational costs and a wider reach. End users similarly can receive broadcasts from everywhere, have more selection, and can use a single technology for both entertainment and data.

It may be some time before Internet radio and television become universal, but when they do, this also opens up the door to liberating more bandwidth for data communications, since broadcast spectrum can eventually be reallocated to IP-based networking.

8.7 Pervasive use

The term pervasive is often used in conjunction with wireless. It means different things to different people, but the common concept is that wireless technology will enable the Internet to expand its reach beyond the office until it permeates our lives. You may wonder whether this is generally desirable, but there are certainly some benefits that can be conceived.

Wireless is a good candidate for the home, not only because it is often easier to implement (no rewiring necessary) but also because it implies access to home information wherever you are.

There are countless potential applications, ranging from home entertainment with a universal remote control (channel changer) to the ability to remotely record a TV program onto a hard disk, download MP3 tracks to a mobile device, or simply watch television over the Internet. Increasingly, wireless technology will also pervade digital cameras and camcorders, making data transfer simpler and faster. Miniaturization will also make it possible to incorporate the technology in watches and jewelry in order to maximize mobility.

It would also be possible to integrate with household heating, surveillance, and lighting systems or remotely control less technical items such as ovens, washing machines, refrigerators, and coffee machines. Some of this is already possible using dedicated wireless spectrum with proprietary encoding. Moving these applications to general-purpose IP networks will again liberate bandwidth, which can then be reallocated to the common pool.

Beyond the home, pervasive use means the ability to interrogate the post office database and determine whether a package has arrived. It should be possible to order a taxi or ask for traffic conditions and directions based on location. Emergency assistance can be facilitated and expedited through wireless connectivity.

8.8 Summary

In a rapidly changing, but still immature, field such as wireless, the future promises many improvements in functionality and these carry with them radically different approaches. Some will succeed and eventually become base technology. Others will fall by the wayside as the market embraces another vision.

We can expect a fundamentally different architecture for mobile applications as the constraints of wireless technology are lifted and software takes careful consideration of the unique needs of mobile environments. The gradual migration to third-generation wireless network will support this new approach by providing higher bandwidth, quality of service, and better mobility.

The applications will also undergo a transformation as they cater to a converged markup language, but the task will not be trivial, since they need to be tailored to a very wide range of machine interfaces and device types. Application development tools will simplify this task by automating much of the diversity, but careful testing and planning will still be necessary.

A wide range of new applications will be enabled by the high bandwidth and better service. In the future it is even possible to envisage a convergence of all kinds of data traffic, including broadcasting and home or personal networking.

All IP also means only IP which means no (or, in the short term, less) need for other spectra—a virtual cycle that improves the quality of data transmission and reduces the need for any other spectra allocations. By releasing currently reserved spectra to general IP traffic it will be possible to combine all data transmission into a more efficient spectrum allocation scheme based on range rather than purpose.

Bibliography

UMTS Forum, Report No. 11 – Enabling UMTS/Third Generation Services and Applications, October 2000.

Stewart, R. et al. Stream Control Transmission Protocol (SCTP) IETF RFC 2960.

Miller, S. *IPv6: The Next Generation Internet Protocol*, Digital Press, 1997.

Kumar, V. et al. *IP Telephony with H.323*. John Wiley & Sons, 2001.

Collins, D. *Carrier Grade Voice Over IP.* McGraw Hill, 2001.

List of Acronyms

AMPS	Advanced Mobile Phone Service
ASIC	Application Specific Integrated Circuit
ASP	Application Service Provider
ASR	Automated Speech Recognition
CA	Certificate Authority
CDMA	Code Division Multiple Access
CDPD	Cellular Digital Packet Data
CHAP	Challenge Handshake Authentication Protocol
CMEA	Cellular Message Encryption Algorithm
CSD	Circuit Switched Data
DECT	Digital Enhanced Cordless Telecommunication
DMZ	Demilitarized Zone
DSP	Digital Signal Processor
DSSS	Direct Sequence Spread Spectrum
EAP	Extensible Authentication Protocol
EDGE	Enhanced Datarates for GSM Evolution
EMR	Electromagnetic Radiation
FDMA	Frequency Division Multiple Access
FEP	Front-end Processor
FHSS	Frequency Hopping Spread Spectrum
GAIT	GSM ANSI-136 Interoperability Team
GAP	Generic Access Protocol

GEO	Geostationary Equatorial Orbit
GPRS	General Packet Radio Service
GPS	Global Positioning System
GSM	Global System for Mobile Communications
HMI	Human Machine Interface
HSCSD	High Speed Circuit Switched Data
IETF	Internet Engineering Task Force
IrDA	Infrared Data Association
IS	Interim Standard
LEO	Low Earth Orbit
LMDS	Local Multipoint Distribution Services
MAG	Mobile Application Gateway
MCM	Multicarrier Modulation
MEO	Medium Earth Orbit
MMC	MultiMedia Consortium
MMDS	Multi-channel Multi-point Distribution System
MS-CHAP	Microsoft – Challenge Handshake Authentication Protocol
MVNO	Mobile Virtual Network Operator
NMT	Nordic Mobile Telephone
OBEX	Object Exchange
OFDM	Orthogonal Frequency Division Multiplexing
OSI	Open System Interconnection
PAN	Personal Area Network
PAP	Password Authentication Protocol
PDA	Personal Digital Assistant
PDC	Personal Digital Cellular
PIN	Personal Identification Number
PKI	Public Key Infrastructure
PLMN	Public Land Mobile Network
PPP	Point-to-Point Protocol

PSTN	Public Switched Telephone Network
QoS	Quality of Service
RADIUS	Remote Authentication Dial-In User Service
R-UIM	Removable User Identity Module
SIM	Subscriber Identity Module
SMS	Short Message Service
SMSC	Short Message Service Center
SWAP	Shared Wireless Access Protocol
TACS	Total Access Communications System
TD-CDMA	Time Division Code Division Multiple Access
TDMA	Time Division Multiple Access
TEM	Telecommunications Equipment Manufacturer
TIA	Telecommunications Industry Association
TTS	Text-to-Speech
UI	User Interface
UIM	User Identity Module
UMTS	Universal Mobile Telecommunications System
USIM	Universal Subscriber Identity Module
VoIP	Voice over IP
VPN	Virtual Private Network
WAG	Wireless Application Gateway
WAP	Wireless Application Protocol
WASP	Wireless Application Service Provider
W-CDMA	Wideband Code Division Multiple Access
WDP	Wireless Datagram Protocol
WEP	Wired Equivalent Privacy
WIM	Wireless Identity Module
WISP	Wireless Internet Service Provider
WLAN	Wireless Local Area Network
WMAN	Wireless Metropolitan Area Network

WML	Wireless Markup Language
WPAN	Wireless Personal Area Network
WSP	Wireless Session Protocol
WTAI	Wireless Telephony Application Interface
WTLS	Wireless Transport Layer Security
WTP	Wireless Transport Protocol
WWAN	Wireless Wide Area Network

Index

2.5G, 49
3G, 210–14
   alternatives, 213–14
   GSM evolution, 211–13
   implementation, 211
4G, 214–15
802 nomenclature, 42–44
802.1x, 159–64
   architecture, 161–62
   authentication, 160
   authentication process, 162
   bridge, 161
   controlled ports, 160–61
   security considerations, 162–63
   use of, 159
   wireless implementation, 163–64
802.11 standard, 42–45
   defined, 42
   development efforts illustration, 44
   evolution of, 44–45
   physical layer operation, 42
   Working Group, 43–44
802.15 standard, 40
802.16 standard, 45

Absorption, 13
Access methods, 122–23
   enhanced, 122
   matching devices to, 126–29
   native, 122
   proprietary solutions, 123
   SMS, 123
   terminal services, 123
   WAP, 123
   Web, 122
Ad hoc networks, 89
Advanced Mobile Phone Service (AMPS), 8, 47
Advertising, 132
Aggregators, 188
Air interfaces, 25–57, 210–17
   3G networks, 210–14
   4G networks, 214–15
   carrier wave considerations, 29–32
   characteristics, 28–38
   current, 38
   data signal considerations, 33–38
   defined, 22–23
   future directions, 210–17
   licensing and, 26–28
   multiple, 91
   range division, 38
   summary, 55–57
   transmission technologies, 215–17
   wireless LANs, 41–45
   wireless MANs, 45–46
   wireless PANs, 39–40
   wireless WANs, 46–55

Air ownership, 25–26
Air security, 156–66
    Bluetooth, 156–58
    WLAN, 158–64, 158–66
    See also Security
Amplifiers, 19
Amplitude Shift Keying (ASK), 33
Antennas
    receiver, 19
    smart, 215–16
    transmitter, 18–19
Applications, 66–70, 117–49
    access methods, 122–23
    advertising, 132
    calendar/contacts, 133
    client-side presentation, 119–20
    components, 119–21
    consumer, 130–32
    CRM, 134–35
    database, 133–34
    development, 69–70, 198–203
    downloading, 204
    electronic books, 68
    electronic mail, 135–44
    enabling, for wireless access, 121–30
    enterprise, 132–44
    entertainment, 68–70
    evolution of, 118–21
    future, 226–29
    games, 69
    initial, criteria for, 129
    instant messaging, 69
    knowledge management portals, 133
    legacy, 118
    market segmentation, 129–30
    mobile-aware, 208–9
    mobile banking, 131–32
    mobile commerce, 131
    multicasting, 228–29
    music, 68–69
    optimized wireless, 145–46

PIMs, 67–68
selection criteria, 123–30
server-side presentation, 120
summary, 149
traditional vs. mobile, 117–18
value, increasing, 147–49
vertical, 144
video, 68
Voice over IP (VoIP), 226–28
See also Mobile devices
Application security, 171–72
Architecture/design, 177–98
Attenuation, 13
Automatic speech recognition (ASR), 77

Bandwidth, 32
    high, 228
    requirements, 229
BlackBerry Desktop Software, 141
BlackBerry Enterprise Server, 141–42
Bluetooth, 39–40, 108–14, 156–58, 196
    ad hoc nature, 157
    AT commands, 111
    audio model, 110
    authentication, 157
    cable replacement, 108–9
    challenges, 109
    core protocols, 109–10
    cryptographic algorithms, 157–58
    defined, 108–9
    device integration, 109
    L2CAP, 110
    LMP, 110
    location attack and, 158
    PIN attack and, 158
    protocol stack, 109–11
    protocol stack illustration, 110
    RFCOMM, 110
    SDP, 110
    smart cards and, 74
    TCS BIN, 111

Bluetooth profiles, 111–14
   Cordless Telephony, 111–12
   Dial-up Networking, 113
   Fax, 113
   File Transfer, 113
   Generic Access, 111
   Generic Object Exchange (GOEP), 113
   Headset, 112
   illustrated, 112
   Intercom, 112
   LAN Access, 113
   Object Push, 113
   Serial Port, 112
   Service Discovery Application, 111
   Synchronization, 113

Calendar/contacts, 133
Carrier wave, 29–32
   bandwidth, 32
   mobility, 31–32
   radio frequencies, 29–30
   range, 30–31
   See also Air interfaces
CdmaOne, 47
Cellular networks, 46–53
   2.5G, 49
   3G, 50
   bandwidth projections, 52–53
   bandwidth variations, 52
   categories, 46
   EDGE, 50
   emerging standards, 48–52
   GPRS, 49–50
   GSM, 51
   IS-95, 52
   IS-136, 51
   mobile telephony, 46, 47–48
   packet data, 46, 48
   PDC, 52
   UMTS, 50–51
   See also Wireless WANs

Certificate Authorities (CAs), 173
Circuit-switched data protocols, 189
Client-side presentation, 119–20
   environments, 120
   in optimized applications, 146
   trade-offs, 120–21
   See also Server-side presentation
Code Division Multiple Access (CDMA), 36, 37, 47
Consumer applications, 130–32
   advertising, 132
   mobile banking, 131–32
   mobile commerce, 131
   See also Applications
Content providers, 196
Cordless Telephony profile, 111–12
Corporate WLANs, 185–87
   access point placement, 187
   ad hoc mode, 185
   infrastructure mode, 185–87
   See also Wireless LANs
CRM applications, 134–35

Database applications, 133–34
   HanDBase, 134
   Oracle Lite, 134
   SQL Server CE, 134
Data signals, 33–38
   circuit vs. packet data, 37–38
   modulation, 33–36
   multiple access techniques, 36–37
   speech coding, 37
   See also Air interfaces
Demodulators, 19
Development, 198–203
   EPOC, 202–3
   Microsoft Mobile Internet Toolkit, 199–200
   Palm OS, 201–2
   platform, 200–203
   tools, 225

Development *(cont'd.)*
  WAP, 199
Device security, 80, 151–56
  password problem, 151–52
  smart cards, 152–55
  virus protection, 155–56
  vulnerable file system, 152
  See also Mobile devices
Dial-up Networking profile, 113
Differential Quadrature Phase Shift Keying (DQPSK), 34
Digital Enhanced Cordless Telecommunications (DECT), 34, 41
Digital Subscriber Line (DSL), 45
Direct Sequence Spread Spectrum (DSSS), 35
Distance barrier, 3–4
Distributed speech processing, 76
Dynamic presentation, 146

Electrically Erasable Programmable Read Only Memory (EEPROM), 153
Electromagnetic radiation, 23–24
  classification of, 15
  defined, 11
  infrared, 16
  light, 15–16
  radio spectrum, 16–17
  spectrum, 15–16
  wave motion, 11–12
  See also Radiation
Electronic books application, 68
Electronic mail, 135–44
  Internet Mail protocols, 143
  Lotus Domino, 142
  Microsoft Exchange, 136–42
  Novell GroupWise, 142–43
  potential improvements, 143–44
  for wireless, 135–36
  See also Applications
Embedded Visual Tools, 201

Enhanced Data Rates for GSM Evolution (EDGE), 50, 212
Enterprise applications, 132–44
  calendar/contacts, 133
  CRM, 134–35
  database, 133–34
  electronic mail, 135–44
  knowledge management portals, 133
  See also Applications
EPOC, 202–3
  applications, 202
  emulator, 203
  PDAs, 65
Extensibility, mobile devices, 73–74
Extensible Authentication Protocol (EAP), 162, 163

Fax profile, 113
File Transfer profile, 113
Fixed networks, 88
Fixed wireless connections, 2
Frameworks, 91–92
Frequency Division Multiple Access (FDMA), 36
Frequency Hopped Spread Spectrum (FHSS), 35
Frequency Shift Keying (FSK), 33
Future directions, 207–31
  air interfaces, 210–17
  applications, 226–29
  development tools, 225
  mobile architectures, 207–10
  networks, 217–24
  pervasive use and, 229–30
  presentation, 224–25
  summary, 230–31

Games, 69
Gaussian Minimum Shift Keying (GMSK), 34

General Packet Radio Service (GPRS), 49–50, 211–12
   roll out, 212
   TDMA support, 50
Generic Access profile, 41, 111
Generic Object Exchange profile (GOEP), 113
Global Positioning Satellite (GPS), 17
Global System for Mobile Communications. *See* GSM
GSM, 34, 47
   algorithms, 165
   defined, 164
   documents, 165
   evolution, 51, 211–13

H.323, 226
HanDBase, 134
Headset profile, 112
Health risks, 10
Heterogeneous networks, 91
High-bandwidth services, 148–49
High-precision services, 148
High-Speed Circuit-Switched Data (HSCSD), 211
Home RF, 41–42
HTML, 198–99
Human-machine interface (HMI), 60–61

iDEN, 48
I-mode, 104–7
   defined, 104–5
   export and, 106–7
   idea behind, 105
   network system, 105
   as proprietary solution, 107
   protocol use, 106
Implementation, 177–205
Information
   recording, 4
   universal repositories of, 209–10

Infrared
   defined, 16
   restriction, 16
   wireless PANs, 39–40
Infrastructure, 85–116
Infrastructure mode, 21
Instant messaging, 69
Interactive Mail Access Protocol (IMAP), 143
Intercom profile, 112
Interference, 13
Internationalization, 197–98
International Mobile Subscriber Identity (IMSI), 164
Internet Mail protocols, 143
IP-based networks, 207–8
IPv6, 221–23
   Address Autoconfiguration, 221
   IPsec as security mechanism, 222
   migration to, 223–24
   mobile, 222
   Mobile IP support, 221
   stack installation, 223
IrDA, 107–8
   defined, 107
   layers, 108
   optional protocols, 108
   support, 107
IS-95, 47
   defined, 47
   evolution, 52
IS-136, 48
   defined, 48
   evolution, 51

Java 2 Micro Edition (J2ME), 65

Keying, 33–34
Knowledge management portals, 133

LAN Access profile, 113
Licensing, 26–28
  issues, 28
  spectrum division, 27
  territorial division, 27
Light, 15–16
Linux, 65
Local Area Networks. *See* Wireless LANs
Local Multipoint Distribution Services (LMDS), 45
Location-based services, 147–48
  high-precision services, 148
  wide area services, 147–48
Long Message Services (LMS), 95
Lotus Domino, 142

Management, 203–4
  challenges, 204
  mobile device, 78–80, 83
  solutions, 203
Manufacturers, 179
Market landscape, 178–79
Market segmentation, 129–30
Markup language convergence, 224–25
MeGaCo, 227–28
Metropolitan Area Networks (MANs). *See* Wireless MANs
Microsoft Exchange, 136–42
  integrated access, 142
  native interface, 136
  optimized access, 136–37
  proprietary interface, 141–42
  SMS interface, 138–39
  synchronized solutions, 140
  WAP interface, 139
  Web interface, 137–38
Microsoft Mobile Explorer (MME), 65
Microsoft Mobile Internet Toolkit, 199–200
Microsoft Outlook Mobile Manager (MOMM), 140

Middleware, 188–92
  corporate scenario, 191, 192
  defined, 188
  functions, 189
Mobile architectures, 207–10
Mobile-aware applications, 208–9
Mobile banking, 131–32
Mobile commerce, 131
Mobile devices, 59–84
  applications, 66–70
  backup and restore, 79
  configurations, 127
  connectivity, 82
  deployment, 79
  embedded and real-time operating systems, 63–64
  extensibility, 73–74
  input, 61, 62
  management, 78–80, 83
  manufacturers, 179
  matching, to access method, 126–29
  from mobile operator, 184
  multi-modality, 81–82
  notebook, 125
  output, 62
  PDAs, 126
  phones, 125–26
  platforms, 63–66
  security, 80, 151–56
  smart cards, 74–75
  summary, 82–84
  support, 80
  synchronization, 70–73, 83
  terminal HMI, 60–61
  tracking and monitoring, 80
  trade-off, 60
  trends in, 81–82
  types of, 60–63
  voice processing, 75–78, 83
Mobile Information Server (MIS), 139

Mobile IP, 218–21
Mobile operators, 181–84
    air interfaces, 183
    coverage, 182
    devices, 184
    pricing, 183
    selection factors, 182
    supplementary services, 184
    support, 183
    transmission quality, 183
Mobile telephony, 46, 47–48
    AMPS, 47
    cdmaOne, 47
    characteristics, 46
    GSM, 47
    iDEN, 48
    NMT, 47
    PDAs, 64–65
    PDC, 48
    phones, 65–66
    standards, 47
    TACS, 47
    TDMA, 48
    See also Cellular networks
Mobility, 31–32
Modulation, 33–36
    keying, 33–34
    spread spectrum, 34–36
    types of, 33
Modulators, 18
Multicasting, 228–29
Multichannel Multipoint Distribution System (MMDS), 45
MultiMedia Consortium (MMC), 41
Multimedia Messaging Services, 95
Multi-modal systems, 81–82
Multi-modal user interaction, 76–77
Multiprotocol Label Switching (MPLS), 218
Multipurpose Internet Mail Extensions (MIME), 143

Multi-vendor network operators (MVNOs), 181
Music application, 68–69

Networked topology, 20–21
Network layers, 21–22
    illustrated, 86
    OSI reference model, 21, 22
Networks, 87–91, 180–84
    3G, 210–14
    ad hoc, 89
    cellular, 46–53
    choice of, 180
    fixed, 88
    heterogeneous, 91
    IP-based, 207–8
    packet data, 37–38, 46, 48
    peer-to-peer, 88
    pure IP, 114–15
    QoS, 217–18
    Ricochet, 90
    virtual private (VPNs), 166–67
    wireless, 88
    WWAN vs. public WLAN, 180–81
Nordic Mobile Telephone (NMT), 47
Notebooks, 125
    configurations, 127
    desktop vs., 125
Novell GroupWise, 142–43

Object Push profile, 113
OpenCard Framework (OCF), 154
Open mobile platforms, 209
Open Systems Interconnection. *See* OSI reference model
Optimized wireless applications, 145–46
    dynamic presentation, 146
    resiliency, 145
    synchronization, 145–46
    See also Applications

Oracle Lite, 134
Orthogonal Frequency Division Multiplexing (OFDM), 215
Oscillators
    receiver, 19
    transmitter, 18
OSI reference model, 21–22
    defined, 21
    illustrated, 22
Ownership and billing, 28

Packet data networks, 37–38
    characteristics, 46
    data transfer rates, 48
    low-speed, 48
Pagers, 8
Palm OS Emulator (POSE), 201–2
Passwords, mobile, 151–52
Peer-to-peer networks, 88
Perimeter security, 171
Personal Area Networks (PANs).
    See Wireless PANs
Personal Digital Assistants (PDAs), 61, 64–65, 82
    configurations, 127
    EPOC, 65
    PalmOS, 65
    PCMCIA slots, 126
    phone convergence, 81
    players, 64
    processing power, 76
    Windows CE, 64–65
Personal Digital Cellular (PDC), 48
    defined, 48
    evolution, 52
Personal Information Managers (PIMs), 67–68, 93
Pervasive use, 229–30
Phase Shift Keying (PSK), 33
Phones, 65–66
    advantages, 126
    configurations, 127
    drawbacks, 125
    PDA convergence, 81
    WAP, 66, 169
Platform Builder, 200–201
Platforms
    development, 200–203
    mobile device, 63–66
    open mobile, 209
Point-to-point wireless connectivity, 8, 20
Portals, 188
Post Office Protocol (POP), 143
Pseudo random code (PRC), 35–36
Public-key infrastructure, 172–74
    certificate revocation, 174
    wireless, 174
    See also Secure transactions; security
Public WLANs, 187–88
Pure IP networks, 114–15
Push Access Protocol (PAP), 225

Quality of Service (QoS), 217–18

Radiation, 10–13, 23
    absorption, 13
    attenuation, 13
    defined, 10
    electromagnetic, 11–12, 15–16
    interference, 13
    mechanical, 10–11
    reflection, 12
    refraction, 12
Radio communication, 14–15
Radio frequencies, 17, 29–30
    applications, 30
    defined, 29
    division, 30
Radio receivers, 19
    amplifier, 19
    antenna, 19

demodulator, 19
oscillator, 19
Radio spectrum, 16–17
Radio transmission, 17–19
    components, 17
    receiver, 19
    transmitter, 18–19
Radio transmitters, 18–19
    antenna, 18–19
    modulator, 18
    oscillator, 18
    transducer, 18
Range, 30–31
Reflection, 12
Refraction, 12
Ricochet network, 90

Satellite communications, 53–55
    defined, 53
    GEOs, 53
    LEOs, 54
    MEOs, 54
    navigation, 54–55
    phones, 55
Secure transactions, 172–74
    enterprise security vs., 172
    public-key infrastructure, 172–74
Security, 151–76
    air, 156–66
    application, 171–72
    Bluetooth, 156–57
    enterprise requirements, 171–72
    mechanisms illustration, 175
    mobile devices, 80, 151–56
    perimeter, 171
    smart cards, 74
    summary, 174–76
    supplementary, 166–70
    virus protection, 155–56
    WLANs, 158–64

Selection criteria, 123–30
    device, 124–29
    network, 123–24
    wireless applications, 129–30
Serial Port profile, 112
Server-side presentation, 120
    environments, 120
    in optimized applications, 146
    trade-offs, 120–21
    See also Client-side presentation
Service Discovery Protocol (SDP), 111
Session Initiation Protocol (SIP), 227
Short Message Service (SMS), 92–96
    applications, 93–94
    Center (SMSC), 94
    future of, 94–95
    interface, 138–39
    lesson from, 95–96
    mechanism illustration, 93
    messages, 92
    modem, 93
    MO messages, 95
    potential enhancements, 95
SIM cards, 154–55
Simple Mail Transfer Protocol (SMTP), 143
Small form factor (SFF) devices, 103
Smart antennas, 215–16
Smart cards, 74–75, 152–55
    EEPROM, 153
    goal, 75
    interfaces, 75
    OpenCard Framework (OCF), 154
    security, 74
    SIM, 154–55
    standards, 153–54
    See also Mobile devices; Security
Software radios, 216–17
Sound, 14–15
    defined, 14
    radio communication vs., 14–15
Speaker verification, 78

Speech coding, 37
Spread spectrum, 34–36
   direct sequence, 35
   frequency hopped, 35
   modulation illustration, 34
SQL Server CE, 134
Stream Control Transmission Protocol (SCTP), 208
Subnet Bandwidth Management (SBM), 218
Subscriber Identity Modules (SIMs), 154–55
Supplementary security, 166–70
Synchronization, 70–73, 83
   commands, 72–73
   in optimized wireless applications, 145–46
   profile, 113
   solutions, 70
   SyncML, 70–73
   types, 71–72
   See also Mobile devices
SyncML, 70–73
   defined, 70
   specifications, 73
   transport protocols, 71

TD-CDMA, 212–13
Telegraph, 5
Telephone, 6
Telephony devices, 63
Text to Speech (TTS), 77–78
Third-Generation Partnership Project (3GPP), 223
Time Division Multiple Access (TDMA), 34, 36, 37, 48, 50
Topology, 184–88
   WAP, 101–2
   wireless, 20–21
Total Access Communications System (TACS), 47
Transducers, 18

Universal Mobile Telecommunications System (UMTS), 50–51
   air interface support, 50
   interfaces, 212
   licenses, 27
   QoS classes, 218
   roll out, 213

Vertical applications, 144
Video application, 68
Virtual private networks (VPNs), 166–67
Virus protection, 155–56
Voice over IP (VoIP), 226–28
   H.323, 226
   high bandwidth, 228
   MeGaCo, 227–28
   SIP, 227
Voice processing, 75–78, 83
   automatic speech recognition (ASR), 77
   distributed speech processing, 76
   multi-modal user interaction, 76–77
   speaker verification, 78
   Text to Speech (TTS), 77–78
   See also Mobile devices
VoiceXML, 76

WAP, 96–104
   architecture, 96–101
   authentication, 168, 169
   competing technologies, 104
   components, 192
   current status, 104
   dedicated dial-in service, 169
   demise, 102–4
   development, 199
   dial-up server requirement, 102
   divergence from Internet standards, 103
   interface, 139
   networks, 192
   phones, 66, 169

poor performance, 104
Push Architecture, 225
schema, 168
sessions, 101
topology, 101–2
WAE, 96
WDP, 100
without WML, 224–25
WML, 97–98
WMLScript, 98–99
WSP, 99
WTAI, 100–101
WTLS, 100
WTP, 99
WAP gateway, 102, 167, 168, 169, 170, 195–96
  choosing, 195–96
  with dial-up services, 194
  hosted on three networks, 194
  public, 193
Wide Area Networks (WANs). *See* Wireless WANs
Wide area services, 147–48
Wideband CDMA (W-CDMA), 50, 212, 213
Windows CE, 200–201
  Embedded Visual Tools, 201
  Platform Builder, 200–201
Wireless Application Environment (WAE), 96
Wireless application gateway providers, 190–92
Wireless Application Protocol. *See* WAP
Wireless Application Service Providers (WASP), 196–97
  defined, 196
  functions, 196–97
Wireless communication
  accessibility, 9–10
  domain of, 7–8
  emergence of, 6–8
  implications of, 22–23

penetration, 9
point-to-point connectivity, 8
wireline communication vs., 7–8
Wireless Datagram Protocol (WDP), 100
Wireless Equivalent privacy (WEP), 158–59, 163
Wireless LANs, 2
  802.1x, 159–64
  802.11, 42–45
  corporate, 185–87
  DECT, 41
  Home RF, 41–42
  PAN contention, 40
  public, 187–88
  security, 158–64
  WEP, 158–59
  WWANs vs., 180–81
Wireless Local Loop (WLL), 45
Wireless MANs, 45–46
Wireless Markup Language (WML), 97–98, 198
  card, 98
  deck, 98
  HTML vs., 97
  WAP without, 224–25
Wireless Metropolitan Area Networks, 2
Wireless networks, 20–21, 24, 88
Wireless PANs, 39–40
  802.15, 40
  Bluetooth, 39–40
  infrared, 39–40
  LAN contention, 40
Wireless physics, 10–17
  electromagnetic spectrum, 15–16
  health and, 10
  radiation, 10–13
  radio spectrum, 16–17
  sound, 14–15
Wireless PKI, 174

Wireless Session Protocol (WSP), 99
Wireless technology, 1–14
   emergence of, 6–7
   excitement, 2
   mobile Internet access and, 2
   summary, 23–24
Wireless Telephony Application Interface (WTAI), 100–101
   defined, 100
   services, 101
Wireless topologies, 20–21
   networked, 20–21
   point-to-point, 20
Wireless Transport Layer Security (WTLS), 100, 167–68
   authentication, 168
   defined, 167
   sessions, 167
Wireless Transport Protocol (WTP), 99

Wireless WANs, 46–55
   cellular networks, 46–53
   network security, 165–66
   public WLANs vs., 180–81
   satellite communications, 53–55
   scenario, 179
   selection, 184
Wireline communication, 4–6
   accessibility, 9–10
   advancement, 8
   constraints, 8–10
   penetration, 9
   telegraph, 5
   telephone, 6
   wireless communication vs., 7–8
WMLScript, 98–99

XTNDConnect Server, 140